Patrick Moore's Practical Astronomy Series

Other Titles in This Series

A Buyer's and User's Guide to Astronomical Telescopes & Binoculars

James Mullaney, F.R.A.S.

 Springer

James Mullaney
Rehoboth Beach
Delaware
USA

British Library Cataloguing in Publication Data
A catalogue record for this book is available from the British Library

Library of Congress Control Number: 2006928725

Patrick Moore's Practical Astronomy Series ISSN 1617-7185
ISBN-10: 1-84628-439-2
ISBN-13: 978-1-846-28439-7

Printed on acid-free paper.

9 8 7 6 5 4 3 2 1

Springer Science+Business Media
springer.com

Preface

In Robert Frost's famous poem *The Star-Splitter,* he states that someone in every town owes it to the town to keep a telescope. I would take that a step further and say that someone in *every home* should have one! For without these wondrous instruments, we are out of touch with the awesome universe in which we live. I have no doubt that Frost himself would have agreed with me, for he was an avid stargazer throughout his long life.

The book you're holding in your hands will make it possible for you to be that someone who has the vision and curiosity to own a telescope. It will help answer such questions as: "Should I buy a new or used telescope?" "Can I make one myself?", "Which type is best?", "What size should I get and how much should I spend?", "How much power do I need?", "What can I see with it once I get it?", and "Do I really need a telescope or will binoculars suffice?" These are all important concerns – and ones that should be addressed before plunging into the purchase of any instrument intended for stargazing.

This book contains two main themes. One deals with the different kinds of astronomical telescopes and binoculars, and recommended sources for them. The other tells you how to use them once you possess them and what to look at in the sky.

Perhaps this is as good a place as any to explain what we mean when we describe an instrument as "astronomical." This term relates to its optical quality. While just about any telescope will show the features of the Moon's alien landscape, the four bright jewel-like satellites of Jupiter, and perhaps even the majestic ice-rings of Saturn, there's an amazing difference in the views seen of these and a host of other celestial wonders through a precision optical system compared with those in one of poor or mediocre quality. Most binoculars and low-end "spotting grade" telescopes are designed with terrestrial use in mind rather than celestial. The optical precision needed to produce razor-sharp views of the Moon and planets and pinpoint images of stars is an order of magnitude above that required for ground-based observing. Since binoculars are normally used at very low magnifications (typically 10× or less), optical aberrations are not nearly as critical an issue for them as they are for a telescope with its correspondingly greater magnifications (typically 50× and higher).

For many readers – particularly those who are already somewhat familiar with telescopes and binoculars, as well as with their use – the greatest value of this book may well prove to be the extensive compilations in Chapter 8 and Chapter 9 that provide handy references to the principal manufacturers and sources for these instruments. Here you'll find standard mail and Internet web-site addresses, telephone numbers (in most cases), and an overview of types and models offered. Since new or upgraded instruments are constantly being added to many manufacturers' lines, you'll be able to get information and specifications on the very latest available models, current prices, and delivery times through their catalogs and other literature (both on-line and printed copies).

Distilled in this volume is the author's more than 50 years of experience in making, designing, selecting, testing – and especially using – literally thousands of different sizes, types, and makes of telescopes. These have included refractors from 2- to 30-inch (!) in aperture, reflectors from 3- to 60-inch, and compound catadioptric scopes from 3.5- to 22-inch in size, employed both for casual personal observing and (in the case of the larger apertures) for research work as well. And after all those instruments and all those years, I'm as excited about telescopes and stargazing as ever!

It should be mentioned here that this present book is *not* intended to be a comprehensive treatise on all the intricate technical aspects of telescope and binocular optics – nor is it intended to be an all-encompassing guidebook on their use in astronomy. (References are given throughout for those who do wish to dig deeper into these areas.) Its purpose, instead, is to offer readers a condensed, trustworthy treatment of these topics that is sufficient to make informed decisions on the selection and use of these instruments – but general enough to not overwhelm them. And although I discuss telescopes costing many thousands of dollars, I also cover ones priced at around a hundred dollars. Stargazing can indeed be a very affordable pastime! If you are new to the field, it's best to start with a basic instrument of good quality (especially optically) and then in time graduate to a larger and/or more sophisticated one if desired. Despite all the varied types and sizes of instruments I've used over the years, my most pleasurable observing experiences still continue to be those with a 3-inch short-focus refractor at 30× and a 5-inch catadioptric telescope at magnifications of 40× to 100×. Each of them, used on an interchangeable, lightweight but sturdy altazimuth mounting with smooth slow-motion controls and wooden tripod, weighs in at less than a dozen pounds. Another long-time favorite instrument is a very portable 6-inch reflector on a basic Dobsonian mounting.

It's my sincere wish that, whatever level your present familiarity and experience with telescopes and binoculars may be, if you read this book carefully you will be able to select a quality optical instrument ideally suited to your particular needs and intended purposes. And more importantly, that it will also lead you to the ultimate use of these marvelous devices – whether binocular or telescope, new or used, large or small, inexpensive and basic, or costly and sophisticated – viewing the wonders of the heavens in a way that will excite, enrich, and ennoble your life, as well as that of others!

James Mullaney
Rehoboth Beach, Delaware
USA

Acknowledgments

There are many people in the astronomical and telescope manufacturing community who have helped to make this book possible. The companies listed in the tables in Chapters 8–9 have kindly supplied the resource information given there on their products, as well as images of selected instruments in many cases. Special thanks must go to Orion Telescopes & Binoculars, which has generously supplied me with images of many telescopes, binoculars, and accessories typical of those widely used by amateur astronomers today.

I am also indebted to Dennis di Cicco, a Senior Editor at *Sky & Telescope* magazine and long recognized as one of the world's most experienced astrophotographers, for the use of previously unpublished images from his private collection. And California astroimager Steve Peters also kindly supplied many of his personal images for use in this book.

Dr John Watson, FRAS, Dr. Mike Inglis, FRAS, and my editors at Springer – Nicholas Wilson in the London office; Dr Harry Blom, Louise Farkas, and Christopher Coughlin in their New York office – have all been most helpful and a sincere pleasure to work with on this, my second volume for this truly world-class publisher.

And finally, I wish to thank my dear wife, Sharon McDonald Mullaney, for her encouragement and continued support of my ongoing mission of "celestial evangelism."

Contents

Part II: **Using Astronomical Telescopes and Binoculars**

Part I

Buying Astronomical Telescopes and Binoculars

CHAPTER ONE

Introduction

More Than Meets the Eye

This book is offered as a no-nonsense practical guide to the selection and use of telescopes and binoculars for stargazing. But these devices should not be looked upon as yet more gadgets to add to our collection of modern technical possessions. Rightly viewed, they are truly magical instruments, for they are literally "spaceships of the mind," "time machines," and "windows on creation" that allow their users to roam the universe in what is surely the next best thing to actually being there!

These lines from William Wordsworth convey something of the excitement that seeing a telescope aimed skyward typically evokes:

> What crowd is this? what have we here! we must not pass it by;
> A Telescope upon its frame, and pointed to the sky.

As you work your way through the many specifications and recommendations contained in the following pages, keep foremost in your mind the wonder of what you're ultimately dealing with in the selection and use of these wonderful devices. To help maintain this perspective, you may want to turn to the concluding chapter from time to time and reflect upon its contents.

New Versus Used Equipment

While this volume focuses on the selection and use of commercially made and available telescopes and binoculars, something should be said about used equipment. Often, a telescope or pair of binoculars can be found on the second-hand used market for a fraction of its original cost new, making it possible to own an instrument that you might otherwise not be able to afford. But the down-side of this is that you have no guarantee of its optical or mechanical condition unless you can actually see and use it before making the purchase. For items bought online over the Internet or by mail, this is not normally possible. In such cases, a substantial deposit should be offered the seller, with the balance to be paid after receiving and inspecting the instrument (and with the clear under-standing that the deposit will be refunded and the instrument returned should any problems be found). The ideal situation is to purchase used equipment within easy driving distance – and preferably from a member of a local astron-omy club – where you can inspect and use it before buying. Aside from examin-

Figure 1.1. Given good optics, even a small telescope can provide a lifetime of celestial viewing pleasure for people of all ages. Shown here is the ubiquitous 2.4-inch (60 mm) refractor, which has long been (and continues to be) the most common telescope in the world. Courtesy of Edmund Scientifics.

ing the tube assembly and mounting for any mechanical damage, you must also carefully check the optical performance using a test like that described later in this chapter.

Among the most sought-after used telescopes are:
    pre-1980 model Unitron refractors (especially the 2.4- and 3-inch)
    Criterion Dynascope reflectors (especially the 6-inch)
    Cave Astrola reflectors (all models)
    Optical Craftsmen reflectors (especially the 8-inch)
    Fecker Celestar (4-inch reflector)
    early models of Questar's 3.5-inch Maksutov-Cassegrain

Mention should also be made of the legendary classic "antique" refractors by such optical masters of the past as Alvan Clark, John Brashear, and Carl Zeiss. With the exception of Unitron and Questar, these firms have been out of the telescope manufacturing business for years, making their instruments true collector's items. If you happen to own one of these gems already – or have an opportunity to purchase one used and in good condition – consider yourself extremely fortunate!

Making Your Own

As with purchasing used equipment, something also needs to be said about the alternative of making a telescope yourself. (Binoculars are not considered here, for their prices are typically so much lower than that of a telescope and their assembly from scratch so much more involved that it's scarcely worth the time and effort to attempt doing this.) And here we need to differentiate between *making* a telescope and *assembling* one. The former involves the time-honored but equally time-consuming art of fabricating the optical components (typically the primary mirror for a reflector and the objective lenses for a refractor). With quality optics mass-produced by machine widely available and reasonably priced today, most "telescope makers" opt for the latter, purchasing the optical components and building the rest of the instrument. This is especially true in the case of the immensely popular, large-aperture Dobsonian reflectors covered in Chapter 5. (A great resource here is Richard Berry's *Build Your Own Telescope*, Willmann-Bell, 2001.) But as a former telescope maker myself, I can attest that there is no thrill quite like viewing the heavens through an instrument that has optics made entirely with your own hands! If you want to go this route, there are many excellent books on grinding, polishing, and testing the mirror for a reflecting telescope. An old standby is *Making Your Own Telescope* by Allyn Thompson, which was reissued by Dover Publications in 2003.

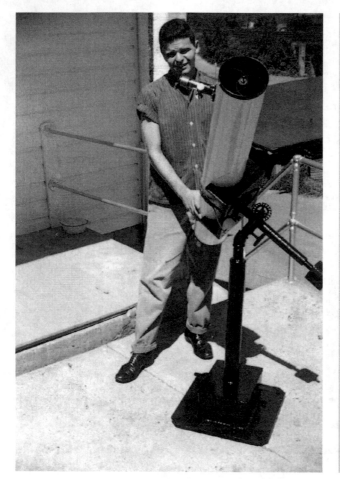

Figure 1.2. The author shown at the age of sixteen with his 6-inch equatorially mounted Newtonian reflector, entirely home-made (including its parabolic primary mirror). Today, most "telescope makers" opt for purchasing commercial optics and mounting them in an instrument of their own construction (typically as a Dobsonian reflector, especially for larger apertures). Photo by the author.

Optical Testing

Whether you purchase a telescope new or used, or make one yourself, you simply *must* know how to test its optical performance! Many sophisticated methods of doing this have been developed over the years by both astronomers and telescope opticians – including Foucault, Ronchi, Hartmann, and interferometric laser testing. But there is one very simple, convenient, and sensitive test that's easy to perform almost anywhere and at any time – even in broad daylight. Known as the *extrafocal image* or *star test*, it uses the image of a star as the test source. This can be either a real one in the night sky or an artificial one produced by shining light through a small pinhole. The latter is especially useful for testing optics in the daytime. (An alternative here is to use the specular reflection of the Sun off the chrome bumper of a car in the distance, or off a glass insulator on a power line; this produces a bright beam of light that is essentially a point source.)

The test is simplicity itself. If you're working with a real star, choose one that's neither too bright nor too faint. An ideal choice is Polaris (α Ursae Minoris), the Pole Star. Not only is it of an ideal brightness but it also offers the great advantage of not moving due to the diurnal rotation of the Earth during testing – a real plus if you're using a telescope without a motor drive! Using a medium-magnification eyepiece (one giving about 20× per inch of aperture), first place the star at the center of the eyepiece field and bring it into sharp focus. Next defocus the star, by going either inside of focus or outside of it, and examine the image. You should see a circular disk within which are concentric rings of equal brightness. (If using a reflector or compound telescope, you will also see the dark silhouette of the secondary mirror at the center of the disk.) Now move an equal distance on the other side of focus. Should you see an identical-looking disk and ring pattern in both positions, congratulations – your telescope has essentially perfect optics! (The technical term for this is *diffraction limited*, meaning that performance is limited solely by the wave nature of light itself rather than the quality of the optical system.)

If optical defects are present, they will readily reveal themselves in the extra-focal image. For example, should the image be triangularly shaped on either side of focus, you have pinched optics. This usually means that either the primary mirror of a reflector or the objective lens of a refractor is mounted too tightly in its cell. This can typically be remedied by loosening the mirror clips in the former case or backing off on the retaining ring in the latter one. If you see an elliptically shaped image that turns 90 degrees as you reverse focus, you have a serious condition known as *astigmatism*. However, before putting the blame on your primary optics, make sure that this isn't in the eyepiece or your own eye! Simply turning the eyepiece in its focusing tube will show if it's the former, while rotating your head will show if it's the latter – in either case, as evidenced by the turning of the ellipse with it.

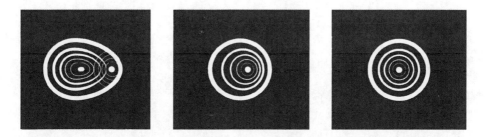

Figure 1.3. The out-of-focus (extrafocal) image of a star can reveal many things about a telescope's optics (as well as about its thermal environment and the state of the atmosphere) – in this case, the alignment of the optical components. The image in the left-hand panel reveals gross misalignment. That in the middle panel shows moderate misalignment (still enough to degrade image quality), while the image in the right-hand panel indicates perfectly collimated optics.

There are other symptoms to look for. Concentric rings that have a jagged or shaggy appearance to them indicate that the optical surface is rough (typically resulting from rapid machine polishing) rather than smooth. Rings that vary in thickness and brightness rather than being uniform in appearance indicate zones (high ridges and low valleys) in the optical surface. Rings that are bunched together and skewed into a comma-shaped image indicate that the optics are misaligned. The extrafocal image can also tell you something about the state of the atmosphere (a rapid rippling across the disk being seen on turbulent nights), the cooling of the optical components (eerie, snake-like plumes moving across the image until the optics reach equilibrium with the night-time air temperature), and the thermal environment of your observing site (waves seen like those rising from a pavement on a warm day).

You should perform this test not only when you buy an instrument but also frequently afterward, in particular to check the optical alignment, or *collimation*. This is especially critical in reflectors and Schmidt-Cassegrains, which can often be thrown out of collimation simply by moving them from place to place. With the exception of well-made refractors and Maksutov-Cassegrain systems (both of which are essentially permanently aligned, owing to the way the optics are mounted in their cells), shipping a telescope is often enough to throw the collimation out. Once learned, the adjustments are relatively simple to perform (especially for a Newtonian reflector) and will make a significant difference in the image quality that you see at the eyepiece.

The finest reference ever written on the subject of extrafocal image testing is Harold Richard Suiter's *Star Testing Astronomical Telescopes* (Willmann-Bell, 1994). It offers an exhaustive treatment of the subject and contains a wonderful array of extrafocal images showing various optical conditions that may be seen at the eyepiece of a telescope. (It should be mentioned here that once binoculars become out of collimation – as evidenced by seeing double images! – they require the services of a professional optician and special alignment jigs to correct, owing to the complex light paths through several trains of prisms.)

CHAPTER TWO

Binoculars

Seeing Double

It's commonly recommended that, before you buy a telescope, you should first get a pair of binoculars. And with very good reason! Not only are such glasses much less expensive, very portable, and always ready for immediate use, but they can also provide views of the heavens unmatched by any telescope. This results primarily from their incredibly wide fields of view – typically 5 or 6 degrees (or 10 to 12 full-Moon diameters) in extent compared with the 1-degree fields of most telescopes, even used at their lowest magnifications. There are also ultra-wide-field models that take in a staggering 10 degrees of sky. Binoculars are ideal for learning your way around the heavens and for exploring what lurks beyond the naked-eye star patterns.

But there's another aspect of "seeing double" (as binocular observing is some-times referred to) that makes these optical gems unsurpassed for stargazing. And that's the remarkable illusion of depth or three-dimensionality that results from viewing with both eyes. This is perhaps most striking when observing the Moon, which through binoculars looks like a huge globe suspended against the starry background – especially during an occultation, when it passes in front of a big, bright star cluster such as the Pleiades or Hyades. See also the discussion in Chapter 13 about apparent depth perception when viewing the Milky Way's massed starclouds with binoculars (or with the unaided eye). Finally, aesthetics aside, it's been repeatedly shown that using both eyes to view celestial objects improves image contrast, resolution, and sensitivity to low light levels by as much as 40%!

Specifications

A binocular consists essentially of two small refracting telescopes mounted side-by-side and in precise parallel optical alignment with each other. Between each of the objective lenses and eyepieces are internal prism assemblies that serve not only to fold and shorten the light path, but also to provide erect images. (Inexpensive "imitation binoculars" such as opera and field glasses use negative eyepiece lenses instead of prisms to give an erect image, resulting in very small fields of view and inferior image quality.)

The spacing between the optical axes of the two halves of a binocular (known as the *interpupilary distance*) can be adjusted for different observers' eyes by rotating the tubes about the supporting connection between them. If this spacing isn't properly set to match the separation between your eyes, you will see two overlapping images. In this same area is a *central focusing* knob that changes the eyepiece focus for both eyes simultaneously. An additional *diopter focus* is provided on most binoculars (typically on the right-side eyepiece) to compensate for any differences in focus between the two eyes. Once this adjustment has been made, you need only use the main focus to get equally sharp images for both. Some lower-grade binoculars offer a rapid-focusing lever; although this allows for quick changes in focus, the adjustment is too coarse for the critical focusing required when viewing celestial objects.

Two numbers are used for the specification of a binocular. The first is the *magnification* or power ($\times$), followed by the *aperture* or size of the objective lenses in millimeters (mm). Thus, a 7×50 glass magnifies the image 7 times and has objectives 50 mm (or 2 inches) in diameter. Another important parameter is the size of the *exit pupil* produced by a binocular, which is easily found by dividing the aperture by the magnification. This means that 7×50 binoculars produce bundles of light exiting the eyepieces that are just over 7 mm across. (These bundles can be seen by holding a binocular against the daytime sky at arm's length. You'll find two circles of light seemingly floating in the air before you.)

The pupil of a fully dark-adapted human eye dilates or opens to about 7 mm, and so in theory all the light a 7×50 collects can fit inside the eye. (This binocular is the famed Navy "night glass" developed long ago by the military for optimum night vision.) But in practice, not only does the eye's ability to open fully decrease with age, but light pollution and/or any surrounding sources of illumination reduce dilation as well. Only under optimum conditions can the full light grasp of a 7×50 be utilized. Thus, a better choice for astronomical use is the 10×50, which gives a 5-mm exit pupil and slightly higher magnification, improving the amount of detail you can see.

A 7×35 or 6×30 binocular also provides a 5-mm pupil, but these smaller sizes have less light-gathering power and resolution than does a larger glass.

Another feature of a binocular to look for is its *eye relief*. This is the distance you need to hold your eyes from the eyepieces to see a fully illuminated field of view. It ranges from less than 12 mm for some models to over 24 mm for others. If the relief is too short, you'll have to "hug" the eyepieces to get a full field of view, and if too long you may have difficulty centering the binoculars over your eyes. A good value is around 15 mm to 20 mm, especially if you wear glasses – if you do, longer eye reliefs are preferred over shorter ones. Note that if you wear glasses simply to correct for near- or far-sightedness (rather than for astigmatism), you can remove them and adjust the focus to compensate. Most binoculars today have fold-down rubber eyecups to enable you to get closer to the eyepieces if necessary; these also keep your eyes from touching the glass surfaces and (depending on style) help keep out stray light.

While just about any size of binocular can be and has been used for stargazing, the 7 × 50 and 10 × 50 are the most popular choices among observers. (See also the section on giant binoculars later in the chapter.) Moreover, 10× and 50 mm are about the highest magnification and largest aperture that can be conveniently held by hand; more power and/or bigger sizes require mounting the binocular on a tripod in order to hold it steady. (It should be mentioned here that *zoom binoculars* are also widely available today. While offering a range of magnifications with the flick of a lever, these generally have inferior image quality and fields of view that change as the power changes.) Good stargazing binoculars in the size range above are available for under $100 from a number of companies, including Bushnell, Celestron, Eagle Optics, Nikon, Oberwerk, Orion, Pentax, and Swift. (See Chapter 8 for contact information on these and many other manufacturers.) Prices for premium astronomical glasses typically run between two and three times this amount.

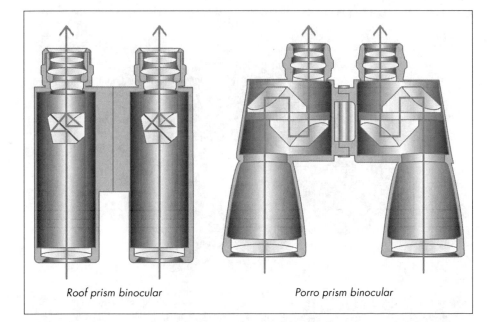

Roof prism binocular   *Porro prism binocular*

Figure 2.1. Optical light path through a roof prism binocular (left) and a Porro prism binocular (right). Although bulkier than the former, the latter is preferred for astronomical viewing because of its superior image quality.

Prism Types and Optical Coatings

There are two basic types of prism assemblies used in quality binoculars today – the more modern and compact *roof prism* style, and the traditional *Porro prism* design. The latter yields images that are brighter and sharper and have better contrast than does the former, but at the expense of more bulk and weight. Porros give binoculars their well-known zigzag shape, while roofs have a straight-through, streamlined appearance. For a variety of optical imaging reasons, Porro prism binoculars are preferred for astronomical use.

Another factor here is the type of glass used to make the prisms themselves. Better-quality binoculars use *BaK-4 barium crown glass*, while less expensive models use *BK-7 borosilicate glass*. BaK-4 prisms transmit more light, producing brighter and sharper images, while BK-7 prisms suffer from light fall-off, resulting in somewhat dimmer images. If the kind of glass used is not stated on the binocular housing itself (where the size, magnification, and field of view are printed), it's easy to find out. Hold the binocular against the daytime sky at arm's length and look at the circles of light (exit pupils) floating behind the eyepieces. BaK-4 prisms produce perfectly round disks, while BK-7 prisms give diamond-shaped ones (squares with rounded corners) having grayish shadows around the edges.

While discussing prism types, mention should also be made of optical coatings. Untreated glass normally reflects 4% of the light falling on it at each surface. Applying *antireflection coatings* (typically magnesium fluoride) to the objective lenses, eyepieces, and prisms can increase light transmission through the binocu-

Figure 2.2. This wide-field 10 × 50 Porro prism binocular is an ideal instrument for general stargazing purposes. Note the coated objective lenses and the cap covering the tripod adapter receptacle located on the bridge joining the two optical barrels. Craters on the Moon, Venus' crescent, Jupiter's four bright Galilean satellites, and awesome views of the Milky Way are just some of the wonders visible through such glasses. Courtesy of Orion Telescopes & Binoculars.

lar significantly. Less expensive binoculars state that they have "coated optics," which normally means that only the outer surfaces of the objective and eyepiece lenses are coated; their inner surfaces and the prism assemblies are not. You can easily check this by looking into the objective end of a binocular and catching the reflection of a bright light or the daytime sky on the glass surfaces. Coated optics typically have a bluish-purple cast to them (this may appear pink if the coating is too thin, and green if too thick), while untreated surfaces will give off white reflections. Quality binoculars specify that they have "fully coated optics," meaning that *all* glass surfaces have antireflection coatings on them. The term "fully multicoated optics" will be found on premium glasses, indicating that several different coating layers have been applied on all glass surfaces to reduce light loss even further. Note that the reflections seen looking into the front of these binoculars typically have a greenish cast to them, mimicking those seen in overly thick coatings as mentioned above.

Image-Stabilized

A fairly recent development introduced by Canon, the well-known camera manu-facturer, is that of *image-stabilized binoculars*. Anyone who has looked through a typical binocular knows at first hand how difficult it is to hold it steady. Even when you are reclining on a lawnchair and supporting both arms, breathing is enough to make the image dance around. (And imagine attempting to use bin-oculars on a rocking boat at sea!). In these "I-S"glasses, roof prism assemblies are essentially floating in sealed oil-filled housings. Microprocessors located within each barrel detect any movement of the observer and send a correcting signal to the prism assemblies to compensate, keeping the image stationary. Available apertures are currently relatively small (10×30 mm to 18×50 mm) and quite costly, with prices beginning around $500 (that of a decent telescope!). Fujinon has also introduced an image-stabilized binocular into its extensive line of high-end glasses. Called the Techno-Stabi, this 14×40 glass is priced at over $1,000. Bushnell and Zeiss are among the companies also now offering image-stabilized binoculars.

Minis and Giants

Binoculars are available in an amazing range of sizes. At the small end are *mini binoculars* – miniature roof prism glasses compact enough to fit in your shirt pocket. Obviously, apertures here are quite limited (typically 25 mm or less in size); although they will show the Moon's surface features, they are not at all suitable for viewing fainter celestial wonders. At the other extreme are *giant binoculars*, with apertures ranging all the way up to 150 mm (6-inch) in size! A giant is generally taken to mean any glass 60 mm or larger in aperture. Among the most common and popular of these are 80-mm binoculars, having magnifica-tions ranging from 11× to 30×. Prices here are much higher than for standard binoculars, running from around $200 to nearly $500. There are exceptions,

Figure 2.3. A 15 × 80 giant binocular for serious two-eyed stargazing! Note, as seen here, that such large glasses must be tripod-mounted, because they are much too heavy to hold steady by hand. Jupiter's disk, the egg-shaped outline of Saturn and its rings, plus hundreds of spectacular deep-sky objects (including the brighter galaxies) lie within reach of giant binoculars. Courtesy of Orion Telescopes & Binoculars.

however. Celestron offers a 15 × 70 glass for under $90, and Oberwerk 8 × 56 and 11 × 56 ones (essentially "giants") for around $100.

There's also the important issue of weight, which for an 80-mm binocular is typically 5 pounds or more. This makes giant glasses all but impossible to hold by hand, requiring them to be mounted on a sturdy tripod. Virtually all binoculars – not just giants – have a provision for adapting them to a tripod mounting. There's typically a cap, located at the objective-end of the central pivot support, covering a standard $\frac{1}{4}$-20 screw receptacle. This takes an L-shaped or "finger" clamp (available from most binocular suppliers) that attaches the binocular directly to the tripod head for support. Before purchasing any binocular, you should carefully check the manufacturer's specifications to see if it is tripod-adaptable. In recent years, sophisticated cantilevered or "parallelogram-style" binocular mounts have also become commercially available, but these can cost as much as a giant glass itself.

Binocular Telescopes

Perhaps the ultimate in giant glasses are *binocular telescopes*. These hybrids are essentially two full-sized telescopes mounted in parallel side by side, with special transfer optics to bring their individual images close enough together to view with both eyes as for a conventional binocular. Initially appearing as home-made curiosities at star parties and telescope-making gatherings (in sizes up to a whopping 17.5 inches in aperture), they were soon followed a few years ago by commercial units introduced by JMI (Jim's Mobile, Inc.) having apertures ranging from 6 to 16 inches. As might be expected since two telescopes are involved, prices here are truly astronomical. A 6-inch binocular telescope goes for around $3,000 and a 10-inch $5,000; larger sizes are essentially custom-made to order and run several times higher.

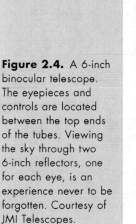

Figure 2.4. A 6-inch binocular telescope. The eyepieces and controls are located between the top ends of the tubes. Viewing the sky through two 6-inch reflectors, one for each eye, is an experience never to be forgotten. Courtesy of JMI Telescopes.

Figure 2.5. A 16-inch giant binocular telescope. The views through dual reflectors of this aperture must be seen to be believed. Many celestial objects appear dramatically suspended three-dimensionally in space (as is the case with binocular viewing in general). Courtesy of JMI Telescopes.

Sources for all the various types of binoculars discussed above (and others) will be found in the comprehensive listing of manufacturers and suppliers in Chapter 8. They should be contacted directly for copies of their latest catalogs, which include detailed specifications and current pricing for all available models. And in conclusion, if you're looking for a good guide devoted entirely to binoculars and their use, an excellent choice is Philip Harrington's *Touring the Universe Through Binoculars* (John Wiley, 1990).

CHAPTER THREE

Telescope Basics

Aperture

The diameter of a telescope's objective (main) lens or primary mirror is known as its *aperture*, which is usually given in inches (and sometimes centimeters) for instruments 4-inch or larger and in millimeters for smaller ones. This is the most important of all a telescope's parameters; for the larger its light-collecting area, the brighter, sharper, and better-contrasted are the images it forms of celestial objects. The primary driving force behind the construction of ever-bigger professional research telescopes (and also that behind the amazing "Dobsonian revolution" sweeping the amateur astronomy community, discussed in Chapter 5) is the need for more light – for collecting ever more photons. (See the discussion on light-gathering power later in this chapter, and also that about the amazing "photon connection" in Chapter 14.) Commercially available telescopes in use by backyard astronomers today range from small 2- and 3-inch aperture refractors up to 36-inch behemoth reflectors, with the most common sizes being in the 4- to 14-inch size range.

Focal Length/Ratio

The distance from a telescope's objective lens or primary mirror to the point where the light it collects comes to a focus is known as its *focal length*. In amateur-class instruments, this is generally stated in inches (and sometimes millimeters for small glasses), but for large observatory telescopes it's often given

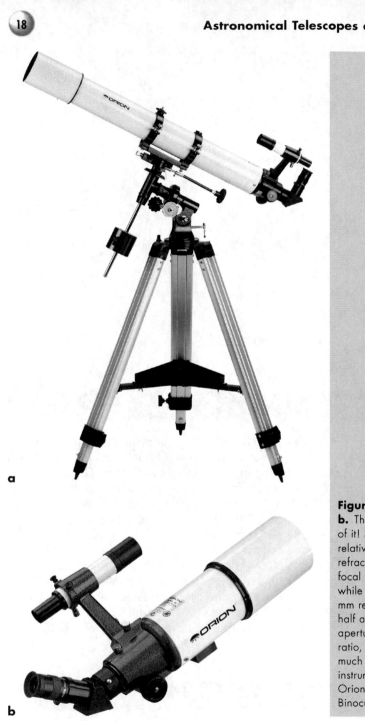

a

b

Figure 3.1a & b. The long and short of it! Shown above is a relatively long 90-mm refractor having a focal ratio of f/10, while below is an 80-mm refractor (less than half an inch smaller in aperture) with an f/5 ratio, resulting in a much more compact instrument. Courtesy of Orion Telescopes & Binoculars.

in feet. The *focal ratio* (or f/ratio) is simply the focal length divided by the aperture (both being measured either in inches or in millimeters). Thus a 5-inch (or 125-mm) telescope with a focal length of 50 inches (or 1250 mm) has a focal ratio of f/10. Similarly, a 6-inch having an f/8 focal ratio has a focal length of 48 inches. Telescopes of f/5 or less are said to be *fast* while *slow* ones are those of f/10 or more. The significance of this will be seen in the section below on magnification and in Chapter 7 when discussing eyepiece fields of view. In general, the faster an optical system is, the more compact it will be and the wider coverage of the sky it provides. On the other hand, the tolerances of its optical surfaces must be significantly higher than for a slow system to form an image of equal quality.

Magnifying Power

While a telescope actually has three different kinds of "powers," that most commonly recognized is its *magnifying power* (×). This is found by dividing the focal length of its objective or primary mirror by that of whatever eyepiece is being used (again, both in inches or millimeters). Thus a 6-inch f/8, having a focal length of 48 inches, when used with an eyepiece with a focal length of 1 inch yields 48×. Similarly, a 5-inch f/10, having a focal length of 1250 mm, when used with a 25-mm eyepiece gives 50×. *Decreasing* the focal length of the eyepiece used *increases* the magnification (because, for example, 1250 mm divided by 10 mm = 125×). Telescopes with longer focal lengths/higher focal ratios yield correspondingly higher magnifications for a given eyepiece over shorter ones. Assuming good optics and a steady sky, the practical upper limit for magnification is around 50× per inch of aperture. On rare occasions when the atmosphere is especially tranquil, as much as 100× per inch may be used on bright objects such as the Moon, planets, and double stars. But it is lower magnifications (7× to 10× per inch of aperture) that typically give the most pleasing results at the eyepiece, owing to their crisp images and wide, bright fields of view. (See the discussion on eyepiece fields in Chapter 7.)

Figure 3.2. As a telescope's magnification is increased, the actual amount of the sky seen decreases (making low powers preferred for many types of observing). Shown here are three views of the Moon at low, medium, and high magnifications. As the image gets bigger, less and less of it can be fitted within the eyepiece field.

Light-Gathering Power

As mentioned above, the most important parameter of a telescope is its aperture or size. This determines its *light-gathering power* – or how bright images will look through it. It's quite important to note here that when the size of a telescope is doubled, it doesn't collect twice as much light but rather *four times* as much, since the area of the optical surface goes up as the *square* of the aperture. Thus, an 8-inch telescope has four times the light-collecting ability of a 4-inch. This means that much fainter objects can be seen through the larger glass than in the smaller one. A 2-inch, for example, will typically show 10th-magnitude stars, while a 16-inch can reveal ones close to 15th magnitude – a factor of 100 times fainter. (See the limiting-magnitude table given in Appendix 1.) Most stargazers start out with a small telescope; but knowing that bigger instruments show more, they often develop what is known as "aperture fever." This is the insatiable desire to own ever-larger telescopes. A limit, of course, must be drawn somewhere. And there's an old maxim that (with the exception of permanently mounted instruments) the smaller the telescope, the more often you will use it. Those of us who have owned more than one telescope can attest to the fact that this is indeed true!

Resolving Power

The ability of a telescope to show fine detail in the image it forms is known as its *resolving power*. This is usually expressed as an angular value in seconds of arc (denoted by ″). There are 60 of these arc-seconds in a minute of arc, and 60 minutes in a degree of sky. At its average distance, the Moon subtends an angle of about half a degree, or 1,800 seconds of arc. So an arc-second is truly a very small angle!

   There are a number of empirical and theoretical criteria used to express the resolving power of a telescope, the best known being *Dawes' Limit*. (See the resolution table in Appendix 1, which gives corresponding values for three of these criteria.) This states that to find out how close in arc-seconds two equally bright points of light (such as those of a matched double star) can be and still be seen as just separated, divide the number 4.56 by the aperture of the telescope in inches. Thus, a 3-inch glass has a resolution of about 1.5″ and a 6-inch 0.76″. This means that the latter telescope can reveal detail twice as fine as can the former one. Note that, unlike light-gathering power, resolving power is a linear relation: doubling the size of the telescope doubles the resolution. Thus, in theory, the larger the telescope, the more detail that will be seen in the image it produces. However, a big limiting factor here is the state of the atmosphere through which we must look. Atmospheric turbulence (or *seeing*, as it's known) makes it difficult to achieve resolutions much under 0.5″ on even the best of nights and more typically less than 1″ in average seeing – no matter how large a telescope you are using. (Professional astronomers today routinely use *adaptive optics* to cancel out the effects of poor seeing, but this highly advanced technology lies far beyond the capabilities and budget of typical backyard astronomers.)

It's interesting to note that one arc-second at the Moon's average distance roughly equals one mile on its surface (its 2,160-mile diameter subtends 1,800 arc-seconds). Thus a 4-inch telescope can see craters and other features a mile or more in size (and considerably smaller if it's a linear feature such as a rill or cleft). Those interested in learning more about telescope resolution should consult the author's book *Double and Multiple Stars and How to Observe Them* (Springer, 2005).

Mountings

There are two basic types of mountings used to support telescopes. Simplest and lightest is the *altazimuth mounting*, which provides a very natural up-down (altitude) and around (azimuth) motion that's easy to use when aiming a telescope at the sky. Heavier and more complex is the *equatorial mounting*, which has one of its axes inclined parallel with the Earth's rotational axis. This makes it possible to compensate for the diurnal movement of celestial objects in the eyepiece by moving the telescope about this axis, either by hand or with a motor drive. Its extra weight over that of an altazimuth largely results from the need to counterbalance the mass of the telescope itself in order to drive it properly. In the common German equatorial mounting, this is done by adding heavy counterweights, while in the fork-style equatorial the telescope is positioned within the fork arms so as to balance itself. The latter form is the one that's most often used on the compact catadioptric telescopes that are so popular today.

In the past, compensating for the Earth's rotation couldn't be done using an altazimuth mount, which must be moved in two directions at the same time to achieve this result. However, with the advent of microprocessors, it has become possible to do this very effectively. Indeed, the famed 200-inch Hale reflector at Palomar was the last of the giant professional telescopes to be mounted equatorially. All the world's great research instruments today – including the huge Keck 400-inch binocular telescope in Hawaii – are computerized altazimuths, greatly reducing the size, weight, and cost of their mountings, as well as of the observatories housing them. The commercial telescope market has quickly followed suit, offering observers the option of light-weight computer-driven altazimuths in place of traditional equatorial mountings.

Whichever type of mounting is selected for the telescope you purchase, you should perform what is known as the "rap or tap test" to check its stability. Simply place a celestial object in the eyepiece and then gently hit the top of the telescope tube with your open palm, noting how long it takes for the image to settle down. A good, stable mounting will dampen its vibrations within a second or two, while a poorly-made, unstable one (such as is commonly found on inexpensive imported refractors) may take 10 seconds or longer! Note that tripods with metal legs typically do not dampen vibrations as quickly, nor as effectively, as do ones with wooden legs. Unfortunately, the latter are becoming harder to find on commercial telescope mountings today. Also, beware of any telescope that uses plastic for the

Figure 3.3. An extremely sturdy, well-made altazimuth fork mounting with wooden tripod legs (which are preferred over metal ones for damping vibrations). Courtesy of Tele Vue Optics.

lens cell or telescope tube – and especially for the focusing mechanism and mounting head. Particularly in cold temperatures, these last two parts bind up and result in frustratingly rough, jerky motions.

The traditional use of mechanical setting circles (subsequently followed by digital ones) displaying Right Ascension and Declination on equatorial mountings to find celestial objects is rapidly disappearing in favor of computerized "Go-To" systems and the truly amazing satellite-based GPS (Global Positioning System). These marvels of technology were introduced by Celestron and Meade, and make it possible to locate thousands of targets essentially at the touch of a few buttons, while providing excellent tracking capabilities. However, for us purists, these devices take much of the fun out of celestial exploration and leave their users not knowing the real sky. An experienced observer with a good star

Figure 3.4. A modern example of a telescope using a massive German equatorial mount – in this case a 14-inch Celestron Schmidt–Cassegrain catadioptric. Like most of today's highly sophisticated mountings, it features automated Go-To finding and tracking capability. Courtesy of Celestron.

atlas and using traditional techniques starting from a bright star and "star-hopping" to the target of interest typically requires less than 10 seconds to find any of a multitude of celestial wonders! Among the several books now available about Go-To systems for those who do want to take advantage of this technology

Figure 3.5. Today's sleek, modern-looking, fork-mounted telescopes offer the convenience of an altazimuth with "equatorial" tracking capability in both axes, plus GPS alignment and acquisition of targets, thanks to the latest computer and aerospace technology. Shown here is Celestron's 9.25-inch-aperture Schmidt–Cassegrain. Its internal database containing over 40,000 objects is typical of those routinely supplied with such premium instruments. Courtesy of Celestron.

is *How to Use a Computerized Telescope* by Michael Covington (Cambridge University Press, 2002). It covers the setup and operation of such instruments by both Celestron and Meade.

From the standpoint of a casual stargazer, I have never minded seeing objects slowly drift across the field of view when I'm using an altazimuth mounting (or an undriven equatorial). This not only provides a vivid demonstration of the rotation of the planet on which we live, but with lines of people waiting to look through the telescope (at a public star party, for example), it naturally limits the viewing time per person! And the slow drifting of objects in the eyepiece – especially faint ones such as nebulae and galaxies – causes their images to move across varying parts of the retina, often revealing subtle details that might otherwise be missed.

Whatever size and type instrument you may be thinking of purchasing, remember that *the best telescope for you is the one you will use the most often.* From the standpoints of light-gathering ability, resolution, magnification, atmospheric turbulence, optical cool-down time, portability, and cost, I believe that a telescope somewhere in the 4- to 8-inch aperture range is perhaps optimum for general stargazing purposes. Many observers wisely opt for having a small portable instrument (often their initial purchase) combined with a much larger but less portable one, giving them the best of both worlds. In this case, a rich-field telescope (or RFT) such as those discussed in Chapters 4 and 5 is a good choice for the former.

Finally, readers looking for a comprehensive all-purpose reference on telescopes and their accessories should consult Philip Harrington's monumental work, *Star Ware* (John Wiley, 2002). The historical aspects of the telescopes covered in the current book have of necessity been kept to a minimum, but it certainly adds to the enjoyment of using them to know something about their fascinating history. Perhaps the ultimate authority here is Henry King's classic *The History of the Telescope*, reissued by Dover Publications in 2003. Those readers particularly interested in classic instruments from the past by the Clarks, Brashear, Fitz, and others master opticians should check out the US-based Antique Telescope Society's web site at http://www1.tecs.com/oldscopes.

CHAPTER FOUR

Refracting Telescopes

Achromatic

The earliest of telescopes was the *refracting* or *lens* type. Initial versions used a single objective lens to collect and focus the image and a simple eyepiece to magnify it. Single lenses by their very nature have a number of optical shortcomings, chief among these being *chromatic* (or color) *aberration*. This results in light of different colors coming to focus at different points, producing image-degrading prismatic haloes appearing around the Moon, planets, and brighter stars. Another serious problem is that of *spherical aberration* – the inability of a single lens to bring all light rays to the same focus. However, as focal length increases, these aberrations tend to decrease. Attempting to take advantage of this, telescopes became ever longer, reaching unwieldy lengths of up to more than 200 feet in one case! (Galileo made his historic discoveries using small, primitive refractors. And while he did not invent the telescope itself, he's credited with being the first person to apply it to celestial observation – and to publish what he saw.)

It was eventually discovered that color aberrations could be considerably minimized by combining two different kinds of glass in the objective, leading to the independent invention of the *achromatic refractor* by Chester Hall and John Dolland. This form is still in wide use to this very day; all the refractors in the world's major observatories (the largest being the 40-inch Clark at Yerkes Observatory) and most of the instruments in amateur hands are basic achromats.

Most common in the latter category is the famed "two point four" or 2.4-inch (60-mm) refractor, with which so many stargazers over the years have begun their celestial explorations. These "department store" telescopes, as they are

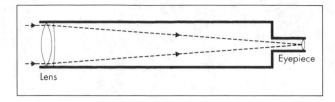

Figure 4.1. The optical configuration and light-path of the classical achromatic refracting telescope, which employs a double-lens objective. (Apochromatic refractors having objectives composed of three or more elements are also widely used by observers today as discussed in the next section.)

sometimes referred to, are mostly imported from Japan (and more recently from China and Taiwan) and typically have optical quality ranging from good to dismal. Even those with good optics often have subdiameter-size eyepieces of poor quality, along with useless finders and shaky mountings. But there are exceptions; the optically and mechanically superb Unitron refractors mentioned in Chapter 1 were and still are made in Japan. If the reader contemplating a first telescope can find a good 2.4-inch or 3-inch Unitron, it can't be beaten for "quick and ready" stargazing at a minimal investment. Among others, Celestron, Edmund, Meade, and Orion offer entry-level 60-mm to 90-mm refractors in the $100 to $300 price range. And fortunately, most companies have at long last switched from the Japanese subdiameter eyepiece size typically found on these imported scopes to the larger and optically superior American standard size (see the section on eyepieces in Chapter 7). Premium-grade achromatic refractors in 4- to 7-inch apertures are offered by Apogee, Borg, D&G, Goto, Helios, Konus, Meade, Murnaghan, Orion, Pacific Telescope, Stellarvue, Unitron, Vixen, Zeiss, and others, with prices ranging from around $500 to more than $2,000.

As is well known, refractors offer the highest resolution (sharpest image) and contrast per unit aperture of all the various types of telescopes. This is because light has an unobstructed path through the system (i.e., there is no image-degrading secondary mirror in the way, as is the case with reflecting and compound systems). They also have very stable images, owing to their closed tube design (eliminating thermal currents within the light path) and rapid cool-down times resulting from their relatively thin objective-lens elements. This has long made them the instrument of choice for high-definition study of the Sun, Moon, planets, and close double stars.

It should be mentioned here that, whatever level of refractor you eventually select, be sure that it has internal *glare stops*. These are concentric flat black rings with a graded series of openings in them, with the largest ones near the front end of the tube and progressively smaller ones toward the eyepiece end as they match the converging cone of light formed by the objective. Their purpose is to eliminate stray light from entering the telescope and reaching the eyepiece. Entry-level refractors generally have two or three glare stops, while premium ones may have five or more. You can check this for yourself by looking into the telescope from the objective end, tilting the tube slightly so you can see down the optical path.

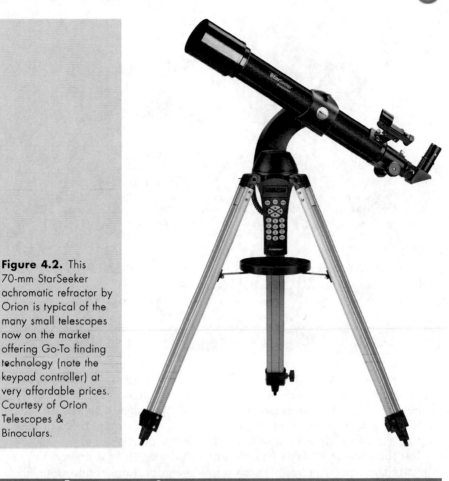

Figure 4.2. This 70-mm StarSeeker achromatic refractor by Orion is typical of the many small telescopes now on the market offering Go-To finding technology (note the keypad controller) at very affordable prices. Courtesy of Orion Telescopes & Binoculars.

Apochromatic

After more than two centuries of widespread use, the traditional achromatic refractor is now giving way to an advanced form called the *apochomatic refractor*. A relatively recent advance that makes use of the latest optical glasses and computerized designs, such an instrument offers excellent correction of all aberrations, including color, by employing three (and in some cases even more) elements in its objective instead of the traditional two. There are also a number of *semi-apochromatic doublets* on the market; these hybrids make use of highly sophisticated rare-earth glasses and optical designs to achieve three-element performance. And while traditional achromats operate at focal ratios around f/11 to f/15 or more in order to control aberrations (making them rather unwieldy in large sizes), apochromats typically work at f/4 to f/6 (and some up to f/9). This results in very compact and highly portable systems. But this superior optical performance and portability come at a hefty price; apochromatic refractors typically begin at around $500 for optical-tube assemblies alone and run into many thousands of dollars for a complete 4- to 7-inch telescope. Among the major

Figure 4.3. The Tele Vue 101-mm apochromatic refractor, seen here equipped with a mirror star-diagonal and Nagler eyepiece. This is one of the finest refracting telescopes ever made, offering superb correction of all possible optical aberrations. These highly color-corrected systems with short focal ratios are mainly available in apertures under 6-inch and are much more costly than a traditional refractor. Courtesy of Tele Vue Optics.

sources for these optically superb instruments are Astro-Physics, Meade, Stellarvue, Takahashi, Tele Vue, and Williams.

Rich-Field

Technically, a *rich-field telescope* (or RFT, as it's referred to) is one that gives the widest possible view of the heavens for its aperture. This occurs at a magnification of about 4× per inch of aperture (so 16× for a 4-inch scope). Because of the relatively long focal lengths of typical non-RFTs, their maximum fields of view are only a degree or so in extent (or two full-Moon diameters of sky) – even when used at their lowest possible powers. But RFTs, with their short focal ratios of typically f/4 to f/5, can easily reach a low magnification, resulting in fields a whopping 3 to 4 degrees across. Sweeping the heavens – particularly the massed starclouds of the Milky Way – with one of these gems is truly an exhilarating experience! RFTs are also wonderful for viewing conjunctions of the planets, big naked-eye star clusters such as the Pleiades (M45), and eclipses of the Moon and (with proper filters!) the Sun. They provide truly amazing low-power views of the Moon itself – especially when it's passing in front of (or occulting) objects behind it, at which times it seems suspended three-dimensionally in space.

Short-focus achromatic refractors are available in apertures ranging from 3 to 6 inches from Apogee, Borg, Celestron, Helios, Konus, Meade, Orion (an especially popular RFT being its little ShortTube 80-mm f/5, priced at $200 without mounting), Pacific Telescope, Stellarvue, and Vixen among others, at prices up to more than $2,000. Many of the shorter aprochromatic refractors offered by several of these same companies, as well as those by Astro-Physics, Takahashi, TBM, Tele Vue, and Williams, also make superb rich-field telescopes – but at prices ranging from under $2,000 to well over $4,000. As with the ShortTube, many RFTs come without a mounting. The achromats are typically lightweight, and at least in the smaller sizes can be supported on a camera tripod, while the apochromats are usually much heavier and require an actual telescope mount. (See also the discussion on rich-field reflecting telescopes in Chapter 5.)

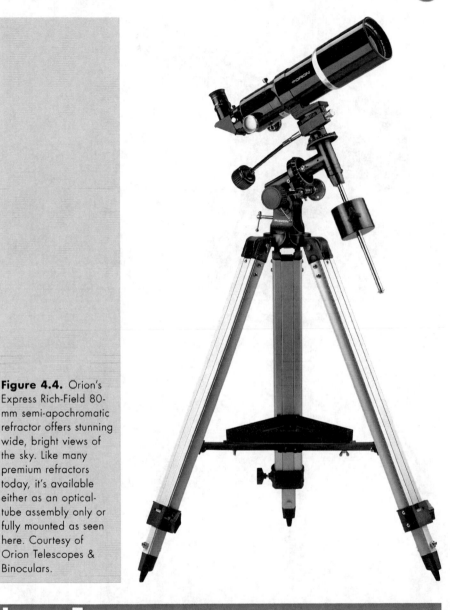

Figure 4.4. Orion's Express Rich-Field 80-mm semi-apochromatic refractor offers stunning wide, bright views of the sky. Like many premium refractors today, it's available either as an optical-tube assembly only or fully mounted as seen here. Courtesy of Orion Telescopes & Binoculars.

Long-Focus

At the opposite extreme from rich-field telescopes are *long-focus refractors*. These are basic achromatic scopes, but have focal ratios from around f/14 to f/20 and sometimes even longer. This results in very large image-scales and high magnifications. These instruments are truly superb for high-definition solar, lunar, and planetary observing, as well as for splitting close double and multiple stars. The views through them when the atmosphere is steady are simply exquisite! (The finest images I have ever seen through *any* telescope have come with the historic 13-inch Fitz–Clark f/14 refractor at the Allegheny Observatory in Pittsburgh,

Figure 4.5. Orion's SkyView Pro 100mm (4-inch) apochomatic refractor on a conventional German equatorial mount. At a focal ratio of f/9, it's one of the longer focal length apos on the market and as such makes an excellent instrument for lunar and planetary observing. Courtesy of Orion Telescopes & Binoculars.

Pennsylvania. In 1870, Samuel Pierpont Langley saw and drew detail in sunspots so fine that it could never be confirmed until balloon-borne telescopes flying in the stratosphere photographed it nearly a century later!)

Unfortunately, these views come with several disadvantages. Long focal lengths and high magnifications bring with them correspondingly limited fields of view – normally much less than a degree, even using wide-angle eyepieces. More of an issue is that these long instruments can become quite unwieldy, requiring very sturdy mountings (and larger structures, if being housed in some type of observatory). Also, with the trend today towards shorter and more portable refractors, sources for such telescopes are relatively limited. A traditional one is Unitron, which offers a 4-inch f/15 achromatic refractor at prices beginning well over $1,000. Another is D&G Optical, which makes 5-inch to 10-inch f/15 scopes on order, with prices running into many thousands of dollars for the larger sizes. Yet-longer focal ratios (ideal for high-resolution lunar, planetary, and double-star observing) must be custom-made, typically at a much higher cost than for a production-line instrument. One solution to the great tube lengths of such refractors is to compress the optical system into only a half or even a third of its original size. This is done by slightly tilting the objective and using one or more optically flat mirrors to fold the light path. While many amateur telescope makers have constructed folded long-focus refractors, to date there have been no commercial versions – or at least ones that have remained on the market for very long! (The ultra-long-focus 24-inch planetary refractor of France's famed Pic du Midi Observatory is an outstanding professional example of folding a telescope's light path.)

It should be mentioned here that, with the use of a 2× or 3× Barlow lens (essentially doubling or tripling the existing focal length – see Chapter 7), a telescope of normal or even short focal ratio can be converted into a much longer one. However, I have never found the image quality to be quite the same as in a true long-focus instrument (owing to a number of subtle optical factors).

Solar

While almost any telescope – but particularly a refractor – can be used for viewing the Sun once equipped with a safe, full- or partial-aperture solar filter placed *over its objective* (*never over its eyepiece!*), there is a type of refractor especially made for solar observation. Not only do these instruments properly filter the Sun, but they also have built-in narrow-bandpass filters tuned to the hydrogen-alpha wavelength. This makes it possible to look into the deeper levels of the Sun's churning atmosphere and also to see graceful prominences dancing in real time off its limb. The premier source here is Coronado, which a few years ago introduced to much acclaim its PST (Personal Solar Telescope) line with apertures ranging from 40 mm to 140 mm in size. These superb instruments provide views of our nearest star that are nothing short of amazing. Prices for the 40 mm begin at about $500. A calcium-light model is now available too. (Coronado also offers its BinoMite 10 × 25 and 12 × 60 filtered binoculars for viewing the Sun in ordinary white light using both eyes.)

CHAPTER FIVE

Reflecting Telescopes

Newtonian

The severe color aberrations of the early single-lens refractors soon led to the invention of the *Newtonian reflector* by Sir Isaac Newton. This form uses a concave parabolic (or spherical) primary mirror to collect light and bring it to a focus. Since the light never passes through the glass mirror but only bounces off its reflecting surface, the image has no spurious color. The converging light-cone reflected from the primary mirror at the bottom of the telescope tube is turned 90 degrees by a small optical flat (or diagonal mirror) before it exits the top and is reflected through the side, where it comes to a focus. All the world's great observatory telescopes today are reflectors of one form or another, including the legendary 200-inch Hale reflector at Palomar and the twin Keck 400-inch reflectors in Hawaii (and the famed Hubble Space Telescope). This is partly because their huge mirrors can be supported from behind (instead of around the edge, as with refractors). It's also due to the fact that the glass itself does not need to be of "optical" quality, since the light merely reflects off its polished and coated surface rather than passing through the glass itself (again, as is the case with refractors).

While reflectors don't suffer from color aberration, they do have a malady known as *coma* – a comet-like flaring of images that is more noticeable further from the center of the field of view. Focal ratios for Newtonians range from f/4 (and as low as f/1 for some professional instruments, to keep them as short as possible) to f/8 or even f/10 in amateur scopes. The shorter the focal ratio, the worse the coma. In the longer lengths, it's hardly noticeable over a typical 1-degree eyepiece field. Also, because of another optical defect known as *spherical aberration* (discussed in Chapter 4 in connection with early refractors), the steeply curving surface of a short-focus primary mirror has to be parabolic in order to bring all

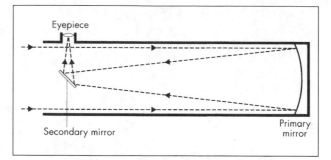

Figure 5.1. The optical configuration and light-path of the classical Newtonian reflecting telescope. A parabolic primary mirror reflects the light onto a small flat secondary one, which directs it to a focus at the side of the tube. Most of the world's large research telescopes are various forms of reflectors.

the reflected light rays to the same focus. But as focal length increases and the curvature of the mirror become flatter, the difference between a paraboloid and a sphere is less distinguishable. Thus, at some point, the mirror may then be left spherical (a much easier optical surface to make) but give an optical performance essentially identical to a parabolic one. It's often stated as a general rule that the focal ratio should be f/10 for this to happen. But the actual value depends on – and increases with – aperture, optical theory requiring less than f/8 for a 4.5-inch, around f/9 for an 8-inch, and just under f/10 itself for a 10-inch.

There's often much discussion about the surface accuracy of telescope mirrors. To satisfy the well-known Rayleigh Criterion for diffraction-limited performance (the point at which image quality is limited by the wave nature of light itself rather than by optical quality), the *wavefront* errors must be $\frac{1}{4}$ of a wavelength of light or less. Since light first enters and then leaves a mirror's surface, surface errors are compounded – meaning that the optics themselves must be figured to at least $\frac{1}{8}$-wavelength to achieve $\frac{1}{4}$-wavelength at focus. (Wavelength accuracy in refractors is rarely mentioned. While in an achromat there are four surfaces – the front and back of each element – and six in a typical apochromat, the light transverses each surface only once. Surface errors averaged over the various elements tend to cancel each other out rather than being compounded. As a result, a lens with surfaces figured to $\frac{1}{4}$-wavelength will still meet the Rayleigh Criterion for total allowable wavefront error.)

Entry-level reflectors are available from a number of manufacturers in 3-, 4-, and 4.5-inch apertures at prices in the $100 to $300 range. Among these are Bushnell, Celestron, Konos, Orion, and Pacific Telescope. Some have parabolic primaries and others spherical ones, in both cases providing acceptable views of the Moon, planets, and brighter deep-sky objects. Unfortunately, they often have less than ideal mountings, which can be quite frustrating to the beginner. As with refractors, a simple, lightweight altazimuth is preferred over a heavier and often clumsy equatorial for casual stargazing. One noteworthy beginner's reflector that has received great reviews in the astronomy magazines is Orion's StarBlast 4.5-inch, which currently sells for around $175. It has an easy-to-use tabletop

Figure 5.2. Orion's AstroView classical 6-inch Newtonian reflector, shown here on a German equatorial mount. The traditional focal ratio for a 6-inch, long preferred by telescope makers and observers, was (and for many still is) f/8. But in a bid to make telescopes more compact and portable, shorter ones have appeared on today's market – as in this case, which is an f/5 system. Courtesy of Orion Telescopes & Binoculars.

Dobsonian mounting (see below) and provides magnifications of 26× and 75× – ideal powers for casual stargazing. While it was designed with young observers in mind, many older amateurs are finding it fun to use as well – especially as a highly portable second telescope. (Its short f/4 focal ratio actually qualifies it as a rich-field telescope or RFT, which is covered below.)

Three very affordable beginner's-level reflectors from the 1950s and 1960s need special mention here. Despite the fact that they are no longer being made, they gave many stargazers their very first views of the heavens, and are still occasionally to be found on the used market today. They were Edmund's 3-inch f/10, Sky Scope's 3.5-inch f/10, and Criterion's 4-inch f/12 Dynascope. All used very basic materials, including bakelite or treated-cardboard tubes and simple but adequate mountings, making it possible to offer reasonably good optical performance at a truly unbeatable price (in the case of the first two scopes just $30, and $50 for the third one!).

Dobsonian

Another type of Newtonian that has become immensely popular with both telescope makers and observers alike in recent years is the *Dobsonian reflector*. Named after the famed "Sidewalk Astronomer" and telescope-maker John Dobson, a Dobsonian is a type of mounting – essentially a basic altazimuth – rather than

Figure 5.3. A fine example of a commercially available Dobsonian reflector – this one the 10-inch Starhopper by Celestron. These affordable instruments offer the most aperture-per-dollar and as such are immensely popular with stargazers today. Courtesy of Celestron.

a form of optical system. Home-made versions typically use simple materials such as plywood for the stand and Teflon strips for the bearings, as well as heavy cardboard tubing for the telescope itself. But there's more to Dobson's brainchild than this. He pioneered the use of very thin mirrors – typically only an inch or two thick and made of plate glass (he originally used ship portholes!) – to make reflectors with very large apertures ranging all the way up to 24 inches. This is a size unheard of in amateur hands until Dobson appeared on the scene. Another distinguishing feature of Dobsonians is their short focal ratios – typically f/4 to f/5, making them very compact for their huge apertures. (Technically, this also qualifies them as rich-field telescopes or RFTs, as discussed below.)

Dobsonian reflectors are now widely available commercially in sizes from as small as 4 inches all the way up to 36 inches! Coulter introduced the first such instrument to the market in 1980 with its 13.1-inch f/4.5 Odyssey-1 for under $500, followed by both smaller and larger models (including a 17.5-inch and a 29-inch!). Sadly, this firm is no longer in business, but many of its bulky but

Figure 5.4. Orion's line of SkyQuest IntelliScope Dobsonian reflectors feature Push-Pull-To computerized technology to locate objects. The hand controller (seen here on the 8-inch model) indicates the location of a desired object while the observer moves the telescope until a "null" reading appears on the LCD display. Courtesy of Orion Telescopes & Binoculars.

economically-priced telescopes remain in use today. Entry level "Dobs" are available from many sources, including Bushnell, Celestron, Hardin, and Orion, in apertures up to 12 inches and with prices starting for under $300 for a 6-inch. Premium Dobsonians are offered by, among others, Discovery, Obsession, and Starsplitter up to 30 inches in size and at prices as high as $5,000. The views of deep-sky wonders such as the Orion Nebula (M42/M43), the Hercules Cluster (M13), and the Andromeda Galaxy (M31) through a large-aperture Dobsonian (even a 10-inch) are truly spectacular, while those in 14-inch and bigger scopes are absolutely breathtaking!

Rich-Field

As with refractors, *rich-field reflectors* provide the widest possible field of view for their aperture. This is again achieved through very short focal ratios (typically f/4), which translate into low magnifications and expansive views. Two

Figure 5.5. The 4.25-inch aperture Edmund Scientifics Astroscan is one of the best-selling small reflectors ever made and it remains one of the most popular RFTs (richest-field-telescope) on the market today. It's 16x eyepiece provides a wide 3-degree field of view (or 6 full-Moon diameters of sky), and its unique optical window protects the system from dust as well as supports its diagonal mirror. Courtesy of Edmund Scientifics.

Figure 5.6. Orion's highly popular 4.5-inch StarBlast reflector, which is a Dobsonian-mounted RFT. While designed for table-top use by young stargazers as seen here, it's also being widely used by seasoned observers as a highly-portable second telescope! Courtesy of Orion Telescopes & Binoculars.

of the best-known such instruments on the market today are Edmund's 4.25-inch f/4.2 Astroscan and Orion's 4.5-inch f/4 StarBlast (discussed above) – both having sturdy tabletop mountings and prices under $200. The former gives a 3-degree field at 16× and the latter about a 2-degree field at 26×, and the optics of both scopes are good enough to support higher magnifications for viewing the Moon and planets. The Astroscan features an optical window that seals the tube and supports the diagonal mirror. Its unique ball-shaped housing sits on an aluminum base with three support pads, creating in essence a very stable universal joint that easily points anywhere in the sky. Both telescopes weigh just 13 pounds, and can be picked up and taken anywhere at a moment's notice. Note, however, that their tabletop mounts do require a sturdy support on which to place them, such as a picnic table. (The Astroscan is tripod-adaptable as well.)

Two other rich-field reflectors deserve mention. If you're on a budget, Bushnell's 4.5-inch f/4.4 Voyager goes for about $130 on a tabletop mount. On the premium end is Parks Optical's 4.5-inch f/5 Companion, which sells for about $800. Among other upgrades included is a tripod altazimuth mount (which brings the total weight up to 30 pounds). Despite the coma inherent in all these short-focus reflecting systems, sweeping the heavens – especially the star-clouds of the Milky Way – and viewing big star clusters such as the Pleiades (M45) or Beehive

(M44) is quite thrilling. Moreover, their low powers are still enough to see the Moon's surface features, the four bright Galilean satellites of Jupiter, and other Solar System wonders such as comets.

Cassegrain

Very soon after the reflecting telescope was invented by Newton, Guillaume Cassegrain introduced a modification now called the *Cassegrain reflector*. Instead of a flat diagonal mirror reflecting the light to the side of the tube, this arrangement substitutes a convex secondary that directs the converging light-cone back down the tube through a hole in the parabolic primary, where it comes to focus. But the light is not just folded upon itself – the secondary's hyperbolic shape changes the angle of the converging beam as if it were coming from much further away. This increases the effective focal ratio of the primary by as much as a factor of five, resulting in an instrument with very long focal length but compressed into a very short tube. This results in high magnifications for viewing (or imaging) the Moon and planets, and for other applications where a large image-scale is desired.

As with refractors, Cassegrains must be carefully light-baffled in order to prevent the field of view from being flooded with light. The glare-stops here are actually carefully machined and fitted flat-black tubing, mounted in front of both the secondary mirror and the central hole in the primary mirror. Having used classical Cassegrains ranging from 6 to 30 inches in aperture, I have never been impressed with their image quality. Not only do they have limited fields of view because of their long effective focal lengths combined with strong field curvature, but the images themselves often seem to have a "softness" about them – certainly not like the crisp, sharp, well-contrasted images of a refractor or a Newtonian reflector.

Many of the optical companies that traditionally offered Cassegrain reflectors commercially to the amateur astronomy market have switched instead to the Ritchey–Chrétien form discussed below. Among the exceptions are Parks, which makes these systems in apertures of 6 to 12.5 inches with prices ranging from $2,000 to $5,000, and Vixen with an 8-inch for under $2,000. Optical Guidance, which is best-known for its Ritchey–Chrétien systems, offers a line of research-grade classical Cassegrains ranging from 10 to 32 inches in aperture with prices beginning around $8,000. Another source is Parallax, which offers a 10-inch at under $5,000, and 12.5-inch and 16-inch models with prices in the $10,000 to $20,000 range.

In a bid to have the best of both worlds, a few manufacturers offer a combined Newtonian–Cassegrain system. An interchangeable or *flip* secondary mirror typically provides f/4 or f/5 wide-field, low-power performance in the Newtonian mode and high-power viewing at f/10 to f/15 or more in the Cassegrain form. These dual instruments tend to be quite pricey, since you're essentially getting two telescopes in one. An example is Takahashi's CN-212 (for Cassegrain–Newtonian), which is an 8.3-inch (21-mm) f/12 Cassegrain with a replaceable secondary mirror that converts it into an f/4 Newtonian, and which goes for over $10,000. Another manufacturer is Parks Optical, which offers f/3.5–f/15 Newtonian–Cassegrains in apertures ranging from 6 to 16 inches at prices from $2,000 to over $16,000. Again, in all cases in which I have used such hybrid instruments, the views at the Newtonian focus were far superior in terms of image sharpness and field of view to the views at the Cassegrain one.

As an aside here, you may wonder why I didn't mention glare-stops and light-baffling when discussing Newtonian reflectors. This is because in a Newtonian the observer is looking into the darkened tube wall across the optical axis rather than along it, as with refractors and Cassegrains. In other words, you are not looking skyward – which is where stray light enters the telescope. However, there's an important point to be borne in mind in this regard when using a Newtonian. As you look into the eyepiece, you are also looking peripherally at the outer surface of the telescope tube as well, which typically has a glossy-white finish. Unfortunately, this makes it an excellent reflector of stray light from around the telescope, thereby reducing the dark adaptation of the eye (see Chapter 10). A flat-black screen made of poster board or other material placed around the focuser/eyepiece area is one solution. Another is to use a photographer's cloth to cover the head and the viewing end of the telescope. Actually, the best color for a telescope tube is red (or no color, i.e. black), which preserves the eye's dark adaptation just as reading star charts with a red light does. (I was a consultant in the development and marketing of Edmund's Astroscan, often referred to as the "red bowling ball." It has an all-red exterior for just this very reason!)

Ritchey–Chrétien

In an effort to improve the imaging quality of the classical Cassegrain reflector for photographic work, George Ritchey and Henri Chrétien jointly developed a marvelous new system in the early 1900s known as the *Ritchey–Chrétien*. Typically operating at effective focal ratios of f/8 or f/9, its hyperbolic primary and secondary mirrors give it both a larger and a flatter field than that of an ordinary Cassegrain, with total freedom from coma. And while it was originally developed for photographic (and more recently CCD) imaging, modernized versions allows for excellent visual observing as well. Initially used by only a few select professional observatories, it has since become *the* system of choice for all major research telescopes built over the past several decades – including the giant twin 400-inch reflectors at the Keck Observatory (currently the world's largest optical telescopes) and the Hubble Space Telescope itself.

Among the first to offer Ritchey–Chrétien reflectors commercially for the serious amateur astronomer and small-observatory market was Optical Guidance, with apertures from 10 to 32 inches, as for its classical Cassegrains. Focal ratios are around f/8 to f/9 and prices range from $10,000 to over $20,000. Its chief competitor is RC Optical, whose line runs from 10 to 20 inches in size in focal ratios of f/8 to f/10, with costs in the same range as those of Optical Guidance. Its popular 12.5-inch f/9, for example, sells for around $14,000. At the time of writing, Meade had just introduced its line of "optically-enhanced" f/8

Ritchey–Chrétien telescopes in apertures from 10 to 16 inches and prices ranging from $6,000 to over $16,000.* These premium telescopes are definitely for the affluent stargazer desiring an observatory-class instrument! Many of the most spectacular photographic and CCD images gracing the pages of the various astronomy magazines such as *Sky & Telescope* over the past several years – some of which rival those taken with large research telescopes – were taken by amateur astronomers using Ritchey–Chrétien systems. And visually, image quality is noticeably improved over that of a classical Cassegrain.

Dall–Kirkham

Another variation on the classical Cassegrain reflector is the Dall–Kirkham, invented by the optician Horace Dall in 1928 and subsequently promoted by the amateur astronomer Allan Kirkham. This form uses an elliptical primary and a spherical secondary mirror, which are easier to figure than those in a standard Cassegrain, thus accounting for its popularity among amateur telescope makers. Its long effective focal ratios (typically f/12 or more) are great for lunar and planetary observing, but the system suffers from coma and noticeable curvature of field. Dall–Kirkhams are not widely available on the commercial telescope market, but one source is Takahashi. Its Mewlon series offers 7-inch, 8-inch, and 10-inch models at hefty prices running between $5,000 and $10,000.

Modified Cassegrain

Several variations on the Cassegrain arrangement have appeared over the years. One is the *coudé* system, widely used by professional observatories in conjunction with both Cassegrain and Ritchey–Chrétien reflectors. Here, a small flat tertiary (third) mirror located above the primary intercepts the converging light-cone from the convex secondary and directs it down the telescope's polar axis. This makes it possible to keep the focal position fixed no matter where the scope is pointed in the sky, and to feed the light into spectrographs and other instrumentation too large and heavy to be supported by the telescope itself. Also, extremely long effective focal lengths (ratios of f/30 or more) can be achieved while keeping the size of the telescope itself manageable.

Another variation is the *modified Cassegrain* form itself. Here, the third flat mirror directs the light out to the side of the tube near the bottom (with a fork mounting, this is usually located right above the balance point where the tube joins the axes). Both the coudé and modified Cassegrain modes eliminate the need to perforate the primary unless desired. This is the system used on the famed 100-inch Hooker reflector at Mt Wilson Observatory, which has no central opening in its huge mirror. By positioning the tertiary mirror in a fork-mounted telescope so that it sends light through and just outside one of the axes themselves, it's possible to keep the eyepiece at a fixed elevation no matter where the instrument is pointed. Invented by James Nasmyth for use on his altazimuth-mounted, speculum-metal-mirrored, 20-inch Newtonian-Cassegrain, this form is referred to as the *Nasmyth focus*. It's being extensively used today on many huge altazimuth, fork-mounted observatory telescopes, including the two 400-inch Kecks, where light is fed into heavy instrumentation located just outside the fork arms. While a number of amateur telescope makers have built such systems in order to have a telescope at which they can remain seated comfortably while observing, currently no commercial coudé or modified Cassegrain telescopes are available other than custom-made ones. The modified Cassegrain form has been employed on catadioptric telescopes (see Chapter 6) as well as on reflectors. Fecker's superb 6-inch f/15 Celestar Maksutov–Cassegrain from the late 1950s was one such instrument, which sold for around $500.

Off-Axis

All early reflectors used mirrors made of speculum metal rather than having reflective coatings on glass, which came much later. In order to reduce light loss from the diagonal mirror in his reflectors, Sir William Herschel tilted their primary mirrors to direct the focus off to the side of the optical path at the top of the tube, where the eyepiece was positioned. This form of unobstructed or *off-axis reflector* is known as the *Herschelian*. Several other designs using tilted primaries and additional flat mirrors to bring the light to a focus without having a secondary in the way have been devised by modern-day amateur telescope makers, one of the more unusual-looking of them being the so-called *Schiefspiegler*. All off-axis instruments must of necessity have long focal ratios (typically at least f/10) in order to tilt their primary mirrors sufficiently for this purpose.

In 2004, Orion introduced a 3.6-inch, f/13.6, modified version of the Herschelian. Instead of having an image that is viewed directly with an eyepiece at the top of the tube, as Sir William did (here quite impractical because of the small aperture), a flat mirror outside the incoming optical path reflects the light across the tube and outside it to a standard focuser. The complete telescope goes for about $1,000 and the optical-tube assembly by itself is available for under $700. Another source is DGM, which offers models ranging from 4- to 9-inch in aperture, with focal ratios averaging around f/10. Their optical arrangement is similar to Orion's, but with the secondary mirror mounted off-axis right under the focuser itself rather than across the tube from it. Prices here range from below $1,000 to over $3,000. Off-axis, unobstructed telescopes offer observers refractor-like performance with the color fidelity of a reflector – but at a significantly higher cost than for a basic Newtonian of similar aperture.

---

* Since the optical enhancement is achieved using a full-aperture corrector plate, technically these systems can be considered catadioptrics.

CHAPTER SIX

Catadioptric Telescopes

Maksutov-Cassegrain

While the inventions of the refractor and reflector occurred within a scant 60 years of each other, no new form of astronomical telescope appeared on the scene for nearly another three centuries. The idea then dawned on telescope designers of combining the attributes of both the refractor and the reflector into a single system, which became known as the *catadioptric* (or compound) *telescope*. In 1930 Bernhard Schmidt used a thin, aspheric corrector plate on a fast Newtonian reflector to flatten and sharpen the field for wide-angle photography, giving birth to the Schmidt camera. A decade later, Dimitri Maksutov combined a thick meniscus lens with a Cassegrain reflector to greatly improve both its visual and its photographic performance, resulting in the *Maksutov-Cassegrain*.

In this system, light entering through the meniscus lens is corrected for the inherent errors resulting from the steep spherical primary mirror. The converging light cone from the primary is then reflected up the tube to a secondary mirror mounted to the back side of the meniscus. In a modification of this scheme known as the *Gregory-Maksutov* and invented by John Gregory in 1957, the secondary mirror is actually an aluminized central spot on the back surface of the meniscus itself. Many instruments marketed today as Maksutov-Cassegrains actually use this system and are, therefore, technically Gregory-Maksutovs.

Introduced by Lawrence Braymer in 1954 after more than a decade of development and testing, the Questar 3.5-inch f/14 Maksutov-Cassegrain became the world's first commercially available catadioptric telescope. (It also holds the record for having the longest continuous production of any telescope in the world – now over half a century!) This exquisite instrument is as much admired

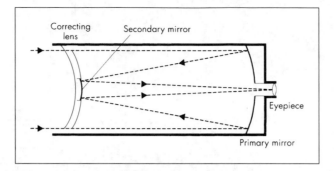

Figure 6.1. The optical configuration and light-path of a catadioptric telescope. The form seen here is the Maksutov–Cassegrain, which uses a thick, steeply-curved meniscus lens to eliminate the aberrations of the steep spherical primary mirror.

Figure 6.2. The legendary Questar 3.5-inch Maksutov–Cassegrain catadioptric, long considered the finest small telescope ever made. This beautiful instrument is truly a work of art, both optically and mechanically. It's seen here in its tabletop altazimuth mode, but it also has legs to tip it into an equatorial position. An engraved star chart (which rotates) on the outer barrel slides forward, serving as a dew cap and revealing an engraved map of the Moon on the telescope's actual barrel. A flip-mirror finder that works through the main eyepiece and a flip-in/out Barlow lens are some of its other unique features. Courtesy of Questar Corporation.

for its beautiful precision-machined tube assembly and tabletop fork mounting as for its unsurpassed optics. Originally priced at about $900, the basic 3.5-inch Questar today goes for over $4,000, making it a telescope mainly for the affluent stargazer and collector. (It can also be found from time to time on the used market for about half this amount.) A 7-inch version and a 12-inch custom-made observatory model are also available at significantly higher prices.

In the 1990s, Meade introduced what is essentially an affordable version of the Questar with its ETX-90 3.5-inch f/13.8 Maksutov–Cassegrain for around $500! This was soon followed by 4-inch f/14 and 5-inch f/15 models. A 7-inch f/15 was eventually added to the line in their LX200 series. The three smaller scopes are all priced under $1,000, while the 7-inch goes for around $3,000. Such economy of pricing for what had traditionally been very costly instruments is a result of Meade's pioneering mass-production of precision-quality optics. The ETX has become so popular that two entire books devoted to its use were published in 2002: *Using the Meade ETX* by Mike Weasner (Springer) and *The ETX Telescope Guide* by Lilian Hobbs (Broadhurst, Clarkson & Fuller).

Figure 6.3. The standard Questar's big brother – a 7-inch Maksutov–Cassegrain. At double the aperture, it has twice the resolution and four times the light-grasp of the smaller instrument, but also much greater cost and weight. Courtesy of Questar Corporation.

Figure 6.4. Both Meade and Orion have introduced their own affordable versions of the pricey Questar Maksutov–Cassegrains at just a fraction of their costs. Seen here is Orion's 127-mm (5-inch) equatorially mounted StarMax catadioptric (which can be purchased as an optical-tube assembly with tripod adapter). It's also available in apertures of 90 mm (the size of the smaller Questar) and 102 mm. Courtesy of Orion Telescopes & Binoculars.

In 2001, Orion entered the field with its StarMax 90-mm f/14 Maksutov–Casseg-rain priced at about $300, followed by 102-mm, 127-mm, and 150-mm models running up to about $700. Among other companies offering these highly popular compact instruments (in apertures all the way up to 16 inches and with prices in excess of $4,000) are Astro-Physics, Orion UK, TEC (Telescope Engineering Company), LOMO (Lenigrad Optical and Mechanical Enterprise), Intes, and Intes Micro. The last three are Russian manufacturers, which seems most appropriate since the Maksutov was originally invented there!

Schmidt–Cassegrain

The most popular and best-known catadioptric system is the *Schmidt–Cassegrain* telescope (or SCT, as it's often referred to). It combines a thin, aspheric Schmidt corrector plate to compensate for the aberrations of a fast, spherical primary mirror in a Cassegrain-style instrument, with the secondary mirror mounted on the back side of the plate itself. Celestron's founder Tom Johnson introduced the first commercial version to the market in 1970. This was the Classic C8 fork-mounted 8-inch f/10 on a sturdy but lightweight field tripod, which eventually replaced the 2.4-inch (60-mm) refractor as the best-selling type of telescope in the world. (There were earlier versions of the Celestron SCT in 10-inch and 16-inch apertures, but these were soon replaced by the C8. They are occasionally still found offered on the used-telescope market today.)

Besides the C8 itself, 5-inch, 9.25-inch, 11-inch, and 14-inch apertures (known as the C5, C9.25, C11, and C14, respectively) were added to the line. (Celestron also made a limited number of 22-inch SCTs for private observatories. The author once spent several nights at a mountaintop observatory stargazing with one of these gems – the views of deep-sky objects through it were nothing short of astounding!) A variety of unique computer-driven, altazimuth, single-arm fork and traditional German equatorial mountings are offered, from basic Go-To systems known as "NexStar" to state-of-the-art GPS systems. Prices today begin under $1,000 for the basic 5-inch, while the 8-inch computerized NexStar goes for around $1,400. The advanced 8-inch GPS model is priced at $2,000 and the 11-inch at under $3,000. Prices for both the 11-inch and 14-inch SCTs mounted on hefty German equatorials begin at well over $3,000, with the top-of-the-line C14 going for nearly $6,000.

In 1980, Meade introduced its own extensive line of Schmidt–Cassegrain tele-scopes, beginning with an 8-inch and eventually followed by 10-, 12-, 14-, and 16-inch models. As with Celestron, these LX200-series instruments are offered with Go-To and GPS capability on computer-driven altazimuth fork mounts. Their Autostar system was actually the very first computerized Go-To system for commercial telescopes. Prices run around $2,300 for the 8-inch, $2,900 for the 10-inch, $3,800 for the 12-inch, and $5,300 for the 14-inch. The 16-inch observa-tory model is offered on either an altazimuth or a German equatorial fork mount and starts at over $15,000.

Following on the immense popularity of Celestron's 8-inch SCT, Criterion introduced its own 8-inch version called the Dynamax, at a lower price than the

Figure 6.5. Today's reincarnated version of the original classic orange-tubed 8-inch Celestron Schmidt–Cassegrain catadioptric, which started the explosive popularity of compound telescopes. It now has such modern features as a sleek single-arm "fork" altazimuth mount and computerized Go-To acquisition of targets and tracking. Courtesy of Celestron.

C8. Bausch and Lomb/Bushnell continued producing this instrument, along with 4- and 6-inch models, when they took over Criterion. Unfortunately, the Dynamax series never gave quite the optical and mechanical performance levels achieved by Celestron (and later Meade) and it was eventually discontinued. These scopes are still to be found today on the used-telescope market, typically at prices far below used Celestron and Meade SCTs. An excellent reference for those considering the purchase of any SCT is *Choosing and Using a Schmidt–Cassegrain Telescope* by Rod Mollise (Springer, 2004).

Schmidt–Newtonian

In an effort to correct for coma in reflectors having a short focal length, the *Schmidt–Newtonian* form was introduced several years ago by Meade in apertures of 6-, 8-, and 10-inch. A Schmidt corrector plate is located at the top of the tube, providing essentially round images right to the edge of the eyepiece field. This plate also seals the telescope tube against dust and thermal currents, and eliminates the need for a secondary mirror support, the mirror being attached to the back side of the corrector itself. These fast systems (f/4 to f/5) give wide, nearly coma-free fields for both visual observing and astroimaging, with prices around $700 for the 6-inch and $1,000 for the 10-inch. A few Newtonians that have also appeared on the market over the years employ an *optical window* to seal the tube and support the secondary, but these have flat (plane-parallel) surfaces and do not provide the optical correction of a Schmidt plate. With the exception of Edmund's Astroscan rich-field telescope (discussed in Chapter 5), there is currently no reflector with an optical window commercially available.

Maksutov–Newtonian

This form of Maksutov combines a steeply curved meniscus, instead of a Schmidt corrector plate, with a fast (typically f/4 to f/6) Newtonian reflector to give superb image quality across a wide field. And unlike the Schmidt–Newtonian, these instruments can also provide detailed views of the Moon and planets. While not nearly as well known as a standard Maksutov–Cassegrain, this form of catadioptric is offered by several companies – three of them from Russia. One of these is LOMO, whose line runs from a 4-inch f/4.5 to an 8-inch f/4.6, with prices ranging from $1,000 to nearly $4,000. Another is Intes, which offers a 6-inch f/6 and a 7-inch f/6 at prices of over $1,000 and more than $2,000, respectively. Intes Micro has a 5-inch f/6 for under $1,000 and a 6-inch f/6 for less than $2,000. Larger models are available in 8-, 10-, 12-, and 16-inch apertures, with prices for the larger sizes running well in excess of $4,000. A fourth, domestic, source of Maksutov–Newtonians is TEC, which offers a 7-inch f/6 and an 8-inch f/3.5 in the $2,000 to $4,000 price range. (In the 1980s, the Canadian firm Ceravolo Optics briefly offered a superb, essentially custom-made, 8.5-inch Maksutov–Newtonian, but did not stay in business for long. These fine instruments had exquisite optics and are much sought after today by observers and collectors.)

Note that many of the prices quoted here are for *optical-tube assemblies* (OTAs) *only*, with the mountings themselves costing extra. You may opt to purchase just the OTA and place it on an existing mount – or perhaps make one, or buy one from another source at a more affordable price (particularly in the case of the three overseas manufacturers). In any and all cases, no matter what make and type of telescope you're interested in, you should contact the companies directly, using the resource information provided in Chapter 9, for specific details on what they are actually offering, availability, current prices, shipping charges, and, of course, delivery time.

CHAPTER SEVEN

Accessories

Eyepieces

It's a telescope's *eyepiece* that does the actual magnifying of the image brought to a focus by the objective lens or primary mirror. It also happens to be the element in the optical train that is most often overlooked as the source of good or bad performance of the overall system. An eyepiece can literally make or break even the best of telescopes. Small imported refractors from the Far East are especially notorious for having poor-quality oculars. And since a telescope would not be able to function without an eyepiece (at least for visual observing), an eyepiece can really be considered a necessity, rather than an accessory as listed here.

There's a multitude of eyepieces on the market today, ranging from inexpensive, basic two-element oculars to sophisticated multi-element designs containing seven or eight individual lenses and costing as much as do some telescopes themselves! A good eyepiece should be well-corrected for chromatic and other aberrations, have as wide and flat (curvature-free) a field as possible, and provide good eye relief. It is especially important that *all* glass surfaces have antireflective coatings to reduce internal reflections. (Some lower-grade eyepieces have so many "ghost" images that they are said to be "haunted"!) In premium eyepieces, the edges of all lens elements are ground and coated flat black in order to eliminate any possible scattering of light. And finally, rubber eyeguards to help position the eye at the correct distance from the eyepiece and also keep out stray light are supplied on most oculars today; if not, they are available separately from many dealers for a variety of ocular sizes, types, and styles.

Eyepieces come in several different sizes of barrel diameters. The 0.965-inch *subdiameter size* (sometimes referred to as the *Japanese size*) ocular is often

Figure 7.1. A fine example of a comprehensive set of quality 1.25″ eyepieces – in this case, Orion's Sirius Plössl collection, having focal lengths ranging from 40 mm to 6.3 mm. For all practical purposes, three eyepieces (providing low, medium, and high magnifications) will suffice for most viewing applications (at least initially!). Courtesy of Orion Telescopes & Binoculars.

found on inexpensive telescopes – especially the ubiquitous 2.4-inch (60-mm) refractor sold everywhere, imported from Japan and other countries in the Far East. They typically have very limited fields of view, poor eye relief, and inferior optical quality. The 1.25-inch *American standard size* is the one most widely used on telescopes, including on many imported scopes in recent years. The larger barrel diameter allows for big multi-element lenses that provide excellent eye relief, roomy fields of view, and good optical corrections. And finally, there's the huge *giant size* 2.0-inch-diameter barrel employed for some of today's most sophisticated, ultra-wide-angle eyepiece designs. They are so big and contain so much glass that they are sometimes referred to as "glass grenades". They also cost as much as do some telescopes!

Of the many types of eyepiece that have been developed over the years, the *Kellner* and the *Erfle* are two of the most common traditionally used by observers. Among the more popular forms today are the *orthoscopic* and the *Plössl*, which not only provide good optical performance and relatively wide fields of view, but are also very reasonably priced. And of the many modern, ultra-wide-field designs now available to stargazers, the *Nagler* series leads the pack with their incredible "space-walk" views (offering up to a whopping 82 degrees of apparent field – see below) and exquisite state-of-the-art optical corrections.

Figure 7.2. One of the legendary Nagler eyepieces having an amazing 82-degree apparent field and providing spectacular "space walk" views of the sky. Courtesy of Tele Vue Optics.

Two basic parameters describe the field of view of an eyepiece. One is its *apparent field* – the angular extent in degrees seen when looking through it at a bright surface such as the daytime sky. This can range from as little as 40 degrees up to as much as the 82 degrees mentioned above, depending on type, design, and brand. Most eyepieces in use today typically have pleasing apparent fields of 50 to 55 degrees. The other parameter is its *actual field* – the amount of sky it encompasses when used on a given telescope. It's quite easy to find what this is: simply divide the apparent field (which is a stated design value for the eyepiece type being used) by the magnification it produces (see below). Thus, an eyepiece having an apparent field of 50 degrees and magnifying 50 times (or 50×) on a particular telescope results in an actual field of 1 degree (or two full-Moon diameters in extent). At 100×, the field becomes $\frac{1}{2}$ degree, and at 200× it shrinks to only $\frac{1}{4}$ degree. Thus, *the higher the power, the smaller the amount of sky a given eyepiece will show*. It should be mentioned here that 1 degree (1°) equals 60 minutes (60′) of arc and that 1′ contains 60 seconds (60″) of arc. The Moon at its average distance has an apparent angular size in the sky of $\frac{1}{2}$ degree or 30′, providing a convenient yardstick for judging eyepiece fields of view.

Determining the magnification an eyepiece gives on a telescope is equally straightforward. The power (×) is found by simply dividing the focal length of the telescope by the focal length of the eyepiece. As already discussed in Chapter

3, the focal length is the distance from a lens or mirror to its focal point, specified in either inches or millimeters. A telescope having a focal length of 50 inches (or 1250 mm) used with a 1-inch (or 25-mm) eyepiece yields a magnification of 50×.

Changing the eyepiece to one with a $\frac{1}{2}$-inch (12.5-mm) focal length increases the power to 100×, while a $\frac{1}{4}$-inch (6-mm) eyepiece gives 200×. Therefore, *the shorter the eyepiece's focal length, the higher the magnification it provides* – and along with it, correspondingly smaller actual fields of view. Thus, it is important to use eyepieces with the largest possible apparent fields. While most telescopes today are typically supplied with one or two basic eyepieces of good quality and reasonable apparent fields of view, you may want to consider eventually upgrading to a premium wide-angle, low-power ocular – one having a focal length, say, between 26 mm and 32 mm.

Most of the major telescope manufacturers and suppliers have extensive lines of eyepiece sizes, types, and designs, ranging from basic oculars priced at under $50 to premium ones going for as much as $300 each! Among others, Orion offers a quality selection of sizes and types at affordable prices, while Meade and (especially) Tele Vue provide premium, multi-element designs in a variety of focal lengths and apparent field sizes at correspondingly higher costs.

Zoom eyepieces make possible a continuous range of magnifications using just a single ocular. These have traditionally been considered much inferior to single eyepieces of a given focal length, owing to the changes in field of view and focus with changes in power. Improved models have recently appeared on the market in an attempt to rectify this. Tele Vue's 8–24-mm Click-Stop Zoom priced at $210 is a definite step up optically over traditional zooms (although the apparent field still varies from 55 degrees to 40 degrees). Its 3–6-mm Nagler Zoom (obviously intended for high-power viewing) has a constant 50-degree apparent field through its short range and sells for around $380. Orion offers a 7–21-mm zoom whose apparent field varies from 43 degrees to 30 degrees; it costs about $60. But for those of us who enjoy wide, expansive eyepiece views, despite their convenience even these improved zooms still fall far short of the performance a high-quality single ocular can provide.

Finders

Another accessory that's often skimped on with a commercial telescope is its *finder*. This is a small auxiliary telescope or other sighting device mounted on the main instrument itself to aid in pointing it at celestial targets so they will appear in the field of a low-power eyepiece. Such an eyepiece typically provides an actual field of only a degree or so; optical finders, on the other hand, have fields of 5 or 6 degrees (similar to those of binoculars), making it easy to locate objects in the sky. Once the finder has been aligned with the main telescope using the provided adjusting screws, so that they're pointing at the same piece of sky, any target placed on the finder's crosshairs will then be in the scope's eyepiece. Magnifications generally range from 6× or 7× for small finders to 10× or 12× for large ones.

Figure 7.3. A conventional straight-through optical finder. Unlike this 9 × 50, finders on many small telescopes are greatly undersized and often are all but useless. Courtesy of Orion Telescopes & Binoculars.

An old rule of thumb states that a finder should have an aperture one-quarter that of the telescope itself. Thus, a 4-inch glass should have a 1-inch finder, an 8-inch a 2-inch one, and a 12-inch a 4-inch one. But this guideline is often ignored by manufacturers in larger-size scopes. And although the guideline may be followed for scopes in the 2–4-inch range, the optical quality is often very poor. A 1-inch (or 25-mm – finder sizes are typically given in millimeters) finder on a 4-inch telescope is hardly adequate. An ideal size for 4–8-inch telescopes is one with a 2-inch (50-mm) aperture and a magnification of 7 times (essentially half of a 7 × 50 binocular!), with correspondingly larger values for bigger scopes. Even a 2-inch glass can benefit from having a finder this size. While tiny finders may be adequate for sighting bright targets such as the Moon and planets, they are nearly useless for locating fainter targets such as nebulae and galaxies.

Having a good finder often requires upgrading the one supplied with the telescope as original equipment – either at the time the scope is ordered from the manufacturer, or as a later purchase from another source. Many of the telescope companies listed in Chapter 9 offer a selection of finders, with prices ranging from as low as $30 to well over $100, depending on aperture. Note that some of these may be equipped with right-angle *star diagonals* (see below) built into them, seemingly to make aiming easier. However, not only do you still have to sight along the tube for initial pointing, but the diagonal produces a mirror-reversed image of the sky that can be confusing to beginners. It should also be mentioned that many of the latest Go-To systems supplied with telescopes today (as discussed in Chapter 3) are so accurate that a finder is not needed. But for quick and ready aiming at bright naked-eye objects, one is still quite useful even so.

In recent years, a new type of finder has been increasingly supplied on telescopes in place of a traditional optical one. Known as the *zero-* or *unit-power finder* (it is actually 1-power – that of the human eye!), this is a sighting device that projects a red dot on the sky as you look through it – typically centered on a bull's-eye pattern. This makes going from a star atlas directly to the sky when aiming a telescope quick, easy, and surprisingly accurate. The original and still one of the best of many such devices now on the market is the famed Telrad, invented by the late Steve Kufeld. If not already included with the instrument you

Figure 7.4. A unit-power (non-magnifying), reflex-sight finder like that now widely used on telescopes in place of (or in conjunction with) conventional optical finders. It works simply by superimposing a tiny LED red dot focused at infinity on a 10-degree view of the sky, showing exactly where the telescope is pointed. Courtesy of Orion Telescopes & Binoculars.

select, these finders can be ordered separately from such companies as Apogee, Celestron, Orion, Photon, Rigel, Stellarvue, Tele Vue, and Telrad itself. Prices range from under $50 to over $100. It turns out that many observers prefer to have *both* a unit-power and an optical finder on their telescopes – the former being used for rapid pointing to the position of a desired target, and the latter for positive identification and precision centering in the eyepiece.

Star Diagonals

Stargazers are typically pictured in cartoons and other media as peering at the heavens straight through a long refracting telescope. This image is quite misleading. For objects on the ground or low over the horizon, this set-up works satisfactorily; but most celestial objects are positioned high in the sky – even at the zenith (or overhead point) – and it's virtually impossible to bend the neck to view them straight through a refractor. This is also true with Cassegrain reflectors and catadioptric systems, where observing is done at the back end of the instrument, as in the case of refractors. (This isn't a concern with a Newtonian reflector, since the observer looks into the side of the tube.) To overcome this problem, a *star diagonal* is used.

Figure 7.5. Shown here is a conventional prism-type star diagonal as commonly used on refractors, Cassegrain reflectors, and catadioptric telescopes. One end fits into the telescope's drawtube and the other end (with lock screw) accepts the eyepiece. Courtesy of Orion Telescopes & Binoculars.

Figure 7.6. An optically perfect mirror star diagonal, which provides better image quality than standard prism diagonals, according to many discerning observers. Courtesy of Tele Vue Optics.

This device consists of two tubes joined at right angles to each other in a housing containing either a precision right-angled prism or a front-surface flat mirror, with one tube fitting into the focuser and the other accepting the eyepiece. The converging beam from the objective or primary mirror is turned 90 degrees to the optical axis by the star diagonal, where the image can be observed in comfort without contorting the neck and back. These are supplied as standard equipment on virtually all refractors and compound telescopes sold today, and are also available separately as accessories from many manufacturers. Prism star diagonals can be had for under $40, while mirror diagonals start as low as $60 and run up into the hundreds of dollars.

It should be mentioned that a star diagonal produces a mirror image of what is being viewed, so objects appear right-side-up but reversed left-to-right. This makes deciding which way to move the eyepiece somewhat confusing until you get used to it. To find your bearings, let the image drift through the eyepiece field (turning off the telescope's motor drive if it has one). Stars will enter the field from the east and leave it to the west. Nudging the scope toward Polaris, the North Star, will indicate which direction is north.

There is another type of diagonal, supplied on some telescopes intended mainly for terrestrial viewing, known as an *erecting prism diagonal*. These diagonals turn the image 45 degrees instead of 90 degrees and provide fully correct images. But not only are they awkward to use for sky viewing because of the angle the light is turned, but the roof prism that erects the image produces an obvious luminous line radiating through bright objects such as planets and first-magnitude stars. As a result, they are definitely not recommended for stargazing purposes!

Barlow Lenses

There exists a marvelous little optical device that effectively doubles or triples the focal length of any telescope, yet measures only a few inches long! Called a *Barlow lens* after the inventor of its optics, it consists of a negatively-curved achromatic lens (sometimes three elements are used instead of two) fitted into a short tube, one end of which accepts the eyepiece while the other goes into the telescope's focuser. With the proliferation of short-focus refractors and fast Dobsonian reflectors in such wide use today, these *focal extenders* are enjoying renewed popularity among observers.

Figure 7.7. This 2× Barlow lens is just 3 inches long and effectively doubles the magnification of any eyepiece used with it. Other models provide amplifications of 2.5× and 3× (or even more, using extender tubes, as mentioned in the text). Courtesy of Orion Telescopes & Binoculars.

The Barlow's negative-lens element decreases the angle of convergence of the light being brought to focus by a telescope's objective lens or primary mirror – causing the latter to appear to be at a much greater distance from the focus than it actually is. This effectively increases the original focal ratio/focal length of the system. Barlows are typically made to amplify between two and three times (2× to 3×). The actual stated *power* is based upon the eyepiece being placed into the drawtube at a set distance from the negative lens; the further the eyepiece is pulled back from this lens, the greater the amplification factor. (Some adjustable Barlows use this very principle to provide a range of powers.) By adding extender tubes, many observers have pushed their 2×- or 3×-rated Barlows to 6× and more! Also note here that the eyepiece–Barlow combination is normally placed into a star diagonal as a unit. But if instead the eyepiece itself is placed into the diagonal and the Barlow inserted ahead of it in the telescope, the extra optical-path length though the diagonal to the eyepiece will also greatly increase its effective amplification.

Solar, lunar, planetary, and double-star observers have long used Barlow lenses to increase the image scale and magnification of the objects they are viewing. The great advantage of these devices to the casual stargazer is that they make it possible to achieve high powers using eyepieces of longer focal length than would normally be required to do so. Such oculars have bigger lenses, wider apparent fields of view, and more comfortable eye relief than do ones of shorter focal length. Thus, an eyepiece having a 25-mm (1-inch) focal length and combined with a 3× Barlow used on a telescope having a 1250-mm (50-inch) focal length would result in a magnification of 150× (3 times 50×). To achieve the same power with an eyepiece alone would require one with a focal length of about 8 mm.

Barlow lenses are not normally supplied as standard equipment on commercially available telescopes (except for imported 2.4-inch/60-mm refractors, which are notoriously already way overpowered without using one!). But they are widely available from many of the companies listed in Chapter 9, at prices from under $50 up to more than $200 for premium units. Here's a great way to effectively double or triple the number of eyepieces in your collection for a very modest investment!

Dew Caps/Light Shields

Reflectors have their own built-in versions of *dew caps/light shields*, since their primary mirrors are located at the bottom ends of their tubes. But refractors and catadioptric telescopes need to have extensions added to their tubes to prevent dew from forming on their front optical elements, and also to help prevent stray light from entering the system. Although refractors are generally provided with a dew cap/light shield, these are typically much too short to offer any real protection. And, surprisingly, virtually every catadioptric telescope sold on the market today comes without one at all! In any case, the observer can (and definitely should) either fashion one out of some black, opaque, flexible material such as common posterboard, or purchase one from the manufacturer at the time the telescope is ordered. They are very affordable (well under $100, depending on size) and are an absolute "must" for anyone using a refracting or catadioptric telescope. (A useful rule of thumb is that a dew cap/light shield should be at least 1.5 times as long as the aperture of the telescope, and to be fully effective 2.5 times as long. The main concern here is that it does not extend out so far as to reduce the effective aperture itself. This can readily be checked by looking up through the instrument without the eyepiece in place at the daytime sky.)

Miscellaneous Items

The following additional accessory items are mentioned here for the sake of completeness. Few are ever supplied as standard equipment with a telescope purchase, and in many cases they have relatively limited utility (especially for beginning observers). In addition to the primary resource listing in Chapter 9, the advertisements in *Sky & Telescope*, *Astronomy* and other magazines provide sources for most of these items.

Binocular viewers make it possible to use both eyes at the telescope instead of one. While some light-loss is involved in splitting the incoming light into two separate beams, as with binoculars image contrast, resolution, color perception, and sensitivity to low light levels are all increased over viewing with one eye only. And there's also the wonderful illusion of depth perception in looking at objects such as, for example, the Moon, where the observer feels suspended in orbit above its vast globe! A downside is the matter of cost. Not only are these devices quite expensive in themselves (ranging anywhere from $300 to $1,600), but two precisely matched eyepieces are needed for each magnification that's used. In other words, a double set of eyepieces is required for the telescope! The binocular viewer fits directly into the drawtube of a Newtonian reflector (it's vital here to make sure the telescope has enough "back-focus" to accommodate the light path through the viewer to the eyepieces; if not, a Barlow lens inserted ahead of the viewer itself can be used to extend the focus), and into the star diagonal of a refractor or catadioptric (which typically have ample back-focus).

Rotary eyepiece holders offer the convenience of having anywhere from three to six eyepieces (depending on model) at your fingertips ready to rotate into position for rapid changes in power. The holder itself is a prism star diagonal and fits directly into the telescope drawtube just as a standard one does. Unitron, with their Unihex rotary eyepiece selector, was the first to market such a device and is still among the few companies offering one. Prices begin around $125. Note that those eyepieces not actually in use often tend to dew up (see below).

Dew heating strips (sometimes called "dew zappers") avoid the annoying formation of moisture on the eye-lens of oculars left exposed to the night air, as well as on objective lenses, corrector plates, and even secondary mirrors. The heating element is typically encased in an elastic nylon strip with Velcro for attaching it around the various optical surfaces and is operated from a 12-volt DC source such as a car battery or power supply. Prices average under $100. While the dew caps discussed above generally provide adequate protection for objective lenses and corrector plates themselves without recourse to heating strips, eyepieces are

Figure 7.8. A binocular eyepiece holder, allowing use of both eyes at the telescope. Note that two oculars of identical focal lengths are required by these units. Some observers actually have a complete double set of eyepieces for use with their bino-viewers! Courtesy of Tele Vue Optics.

particularly vulnerable to dewing up. (So too are the lenses on finders.) They should never be left exposed to the night air in an open eyepiece box, for example. Except for the ocular that's in use on the scope at the time, they should be kept covered. Note that rotary eyepiece holders do leave their eyepieces exposed to the air, often requiring that they be capped until positioned into place for viewing. (Many observers today also employ ordinary hairdryers to remove dew from the various optical surfaces of their telescopes, but care must be taken not to overheat them! In this case, dew is dealt with after it forms on the optics – heating strips prevent it from forming in the first place.)

Image erectors are typically supplied with small imported refractors for use in terrestrial viewing. Their long tubes make them awkward to use on a telescope and their optics often leave much to be desired. An image-erecting star diagonal (mentioned above) offers a much more convenient and optically superior way to achieve a fully corrected image for land-gazing.

Focal reducers can be thought of as "reverse Barlows" in that they reduce the effective focal length of a telescope rather than extend it. Originally developed for use on catadioptric systems with their long focal ratios (typically f/10 to f/14), they're intended primarily for increasing the photographic speed of these slow telescopes for astroimaging purposes by reducing their effective ratios by as much as half the original values. This correspondingly both reduces the lowest achievable magnification and increases the maximum actual field of view that can be obtained with a given telescope. However, focal reducers have seemingly found only limited use for visual work among stargazers.

Aperture masks are used to reduce the effective aperture of a telescope, which many observers feel improves the visual image quality and reduces image motion under conditions of less than ideal atmospheric seeing. This goes along with the claim that small apertures are less affected by poor seeing – supposedly because the turbulence "cells" average around 4–6 inches in width, so that only one or two are over a small telescope at any given instant, whereas many may be over the light-collecting area of a large telescope. Although reducing the aperture can indeed often improve image quality on the Sun, Moon, planets, and double stars in poor seeing, it reduces the resolution and light-gathering power of the telescope as well. The masks can be made from a piece of cardboard simply by cutting a hole smaller than the original aperture itself.

Note that the opening should be on-axis in the case of refractors, and off-axis for reflectors or catadioptrics in order to avoid their central obstructions (which limit the mask's clear aperture to less than the radius of the primary mirror). They are placed over the front of the telescope itself.

Coma correctors do just what the name implies – reduce the amount of coma in fast (f/3 to f/6) short-focus Newtonian telescopes. This is especially useful for the immensely popular large Dobsonian reflectors in use today, most of whose parabolic mirrors operate at f/4.5 and exhibit noticeable flaring of images a short distance from the center of the eyepiece field. Few of these devices are found commercially at present – Tele Vue's Paracorr corrector is one of them and probably the best ever made. As with some of this company's famed wide-angle Nagler series of eyepieces discussed above, its coma corrector costs as much as a small introductory telescope itself! But the improvement in image quality and useable field of view are well worth the price for those who can afford this accessory.

Photographer's cloths are simply dark, opaque pieces of fabric that are thrown over the observer's head and the eyepiece area of the telescope to eliminate stay light and preserve dark adaptation (see Chapter 10). They are available commercially from camera stores and some telescope dealers, and are also easily made. In practice, these can prove a bit suffocating, especially on warm muggy nights, and are sure to raise the eyebrows of any neighbor who happens to see you lurking in the dark!

Telescope covers are used to protect a telescope's sensitive optics (and its mounting) from dust, pollen, moisture, and other airborne contaminants at all times when the telescope is not in use. While they may simply consist of plastic sheeting thrown over the entire instrument, more typically they consist of a fitted plastic cap supplied with the telescope for covering both ends of the tube, in the case of a reflector, or the objective lens or correcting plate for refractors and catadioptrics, as well as eyepieces and finders. The best way to keep a telescope clean is not to let it get dirty! If a cover is not already supplied as standard equipment, a plastic-bowl cover or a heavy-duty shower cap can also be used for this purpose.

Filters of many different types and intended purposes are offered commercially for use on telescopes today. Among these are solar, lunar, planetary, nebula, and light-pollution filters.

With the exception of solar filters, which are placed over the front of the telescope (*never* over an eyepiece, as implied with those supplied on many small imported refractors), the other types screw into the front end of standard eyepiece barrels – virtually all of which today are specifically threaded to take them. I have never been a big fan of filters (except, of course, ones for viewing the Sun!), but they do serve a purpose. Planetary observers have long used them to enhance surface or atmospheric features, and an entire set of them can be purchased for as little as $50. Many deep-sky observers today routinely use nebula and light-pollution filters to increase the visibility of faint objects. These are much more specialized and difficult to manufacture than are planetary filters, with single units beginning at $50 and up. It's perhaps best to use your new telescope for a while to see where your interests lie before investing in them. Full-aperture optical-glass solar filters run from about $60 to nearly $150, depending on aperture, and are an absolute *must* for observing the Sun. Less expensive but still safe Mylar® versions are also widely used.

Micrometers are devices for measuring the angular size or separation of celestial objects (usually in arc-seconds) and their relative positions (or position angle) on the compass heading in degrees with a telescope. Of the many different types in use, the filar micrometer is the traditional such device. The few available commercial models run from around $600 to several thousand dollars, depending on features (such as digital readouts). Another form that's becoming more popular today and is much more affordable is the reticle eyepiece micrometer, which costs about the same as a good eyepiece. Micrometers are most often used in measuring the separations and position angles of double stars (especially binary systems), an activity ideally suited to amateurs looking for a useful observing program to pursue. For more information about micrometers and their application to double stars, see *Observing and Measuring Visual Double Stars* by Bob Argyle (2004) and my own *Double and Multiple Stars and How to Observe Them* (2005), both published by Springer.

Photometers measure the apparent brightness or magnitude of celestial objects (particularly stars) in visual or other wavelengths, generally employing sensitive photocells and electronic circuitry. For the amateur astronomer, they find most application in following the changes in brightness of variable stars. Commercial units are few and far between, and as a result many observers have built their own devices.

Spectroscopes use one or more prisms or a finely ruled diffraction grating to separate the light from celestial objects into its component colors or wavelengths. This makes it possible to glean such amazing physical information about them as their temperatures, compositions, sizes, and rotational and space velocities. For amateur use, the fun is seeing the absorption lines and bands in the various spectral classes of stars. In years past, Edmund marketed an imported eyepiece spectroscope that became very popular and is still to be found on the used market. Today, Rainbow Optics among a few others offers a visual star spectroscope that fits over a standard eyepiece – one capable of showing not only the absorption lines and bands in the brighter stars but also emission lines if present. The visual model sells for $200 while one that includes photographic and CCD imaging capability as well goes for $250. Readers interested in learning more about visual spectroscopes and stellar spectroscopy should consult Mike Inglis' excellent book *Observer's Guide to Stellar Evolution* (Springer, 2003).

Astrocameras come in many different types and varieties, ranging from basic 35-mm film cameras riding piggyback on telescopes for wide-angle shots of the sky to special cameras designed for prime-focus or eyepiece-projection photography through the telescope itself. (Unitron in its early days offered a superb astrocamera for these latter forms, which can still occasionally be found on the used market.) This is a vast and complex field – yet one that can be very rewarding, given lots of patience and practice. Amateur astronomers today are routinely taking spectacular color images of celestial wonders that rival those from the large professional observatories themselves. Even common digital cameras are now being widely used to do astro-imaging through telescopes. But film photography itself is a field that's rapidly declining in favor of CCD and video imaging (see below). A plethora of practical guides covering both conventional and electronic astroimaging are available today for the amateur astronomer. An excellent source for many of these is Sky Publishing Corporation's catalog, available by mail or on-line at www.skyandtelescope.com. Pricing for cameras intended for sky-shooting is about the same as for those used in conventional ground-based photography, since they are essentially the same equipment. But here I should like to offer a word of advice. If you are new to astronomy, before plunging into astroimaging of whatever type, spend the better part of a year seeing the real sky – that of all four seasons – with your own eyes rather than that of a camera! (And in this regard, see the discussion concerning the "photon connection" in Chapter 14.)

CCD and video imagers use charge-coupled devices (CCDs) and either eyepiece video cameras or common webcams, respectively, to photograph the heavens through telescopes. CCD imaging in particular has virtually replaced conventional film photography at most of the world's major research observatories today owing to its immensely faster speed (or "quantum efficiency") and dynamic range, and the fact that the images collected can be immediately viewed and processed electronically using sophisticated computer software. Exposure times of minutes or even seconds now show what previously took hours employing the fastest films. And these devices have now become widely available to amateurs as well. Basic units are surprisingly affordable; Orion offers a black and white Electronic Imaging Eyepiece camera that displays pictures from the telescope directly onto a TV screen, VCR, or camcorder for $65 and a color version for $120.

Meade pioneered affordable CCD imaging systems for their telescopes, including the very popular and easy-to-use Deep Sky Imager with Autostar Suite processing software that sells for just $300 and promises successful images the first night out! But here again, see the advice given in the section above on astrocameras.

Computers have become important tools in observational astronomy, as in almost every other area of modern life. While they can hardly be considered an "accessory" for the telescope in the normal sense of the word, they are used for such tasks as helping to find and track celestial objects, and in making, processing, and displaying observations by electronic imaging – all typically done remotely from the observer's living room, den, or office. While all this certainly has its place, such "robotic" remote observing, however satisfying and comfortable (especially in muggy or frigid weather), is *not* seeing the real sky – and is often not even being out *under* the real sky! Once more, to gain a perspective on this issue, please see the discussion concerning the "photon connection" in Chapter 14.

Figure 7.9. Electronic eyepiece cameras (both black & white, and color) like that seen here make video imaging easy and affordable today. The camera output can be displayed in real time on a monitor, or recorded on a VCR or camcorder for viewing later. Courtesy of Orion Telescopes & Binoculars.

Figure 7.10. Shown here is a CCD color imaging camera attached to the focuser of a Newtonian reflector. This one is intended primarily for use on deep-sky objects, while other models are available for imaging solar system targets such as the Moon and planets. In either case, the output is fed into a PC for viewing and processing. These state-of-the-art devices make it possible for amateur astronomers to routinely take pictures rivaling those of professional observatories! Courtesy of Orion telescopes & Binoculars.

Setting circles/Go-To/Push-Pull-To/GPS systems are all devices designed to help the observer find celestial objects, employing various levels of sophistication. Some are included on certain telescopes as standard equipment, while in other cases they are add-ons to be ordered along with the telescope itself. As already mentioned in Chapter 3, the traditional use of mechanical setting circles (and subsequently digital ones) on equatorial mountings displaying Right Ascension and Declination to find celestial targets is rapidly disappearing in favor of these state-of-the-art computerized systems. These make it possible (after initial setting on two or three bright alignment stars) to locate thousands of objects essentially at the touch of a few buttons while at the same time providing excellent tracking capabilities. In the case of Push-Pull-To systems (as offered by Orion on its IntelliScope series of Dobsonian reflectors), after the target name or designation is entered on the keypad, the observer moves the telescope by hand (instead of with drive motors) until a "null" or zero reading is displayed on the LCD display. The object will then be in the eyepiece's field of view. These devices make finding things relatively easy and are especially helpful under light-polluted skies or when there's little time available to search for elusive targets. They certainly do serve a purpose and are firmly entrenched in modern-day amateur astronomy. Again, however, for us purists, automated acquisition takes much of the fun out of celestial exploration and typically leaves the observer not knowing the sky. We prefer old-fashioned, leisurely star-hopping from bright naked-eye stars to the object sought after using a good star atlas – enjoying the many delightful and unexpected sights encountered along the way!

CHAPTER EIGHT

Binocular Sources

Product and Contact Information for Principal Manufacturers/Suppliers

Presented here in alphabetical order is a listing of the primary sources for astronomical binoculars. Note that this compilation does *not* include dealers themselves, advertisements for whom can be readily found in the various astronomical magazines such as *Sky & Telescope*. Under the company's name are its mailing address, phone number/s where available, Internet web site and/or e-mail address, and a concise listing of those glasses being offered at the time of writing. Apertures smaller than 30 mm are not included here, being largely unsuited for astronomical use. Readers should contact these sources directly for copies of their latest catalogs, specifications for specific models, current prices, shipping charges, delivery times, and warranty and return policy. In some cases, their catalogs are also available on-line as well as by mail. Porro and roof-prism styles, zoom and waterproof models, wide-angle and ultra-wide-angle glasses, mini and giant binoculars, affordable basic units and costly premium ones (including image-stabilized glasses and even telescopic binoculars) are all to be found among the various product lines offered by the sources below!

Apogee, Incorporated
  P.O. Box 136, Union, IL 60180
  Phones: 815-568-2880/877-923-1602
  Web site: www.apogeeinc.com E-mail: apogee@sbcglobal.net

Models: 7 × 50, 10 × 60, 12 × 60, 20 × 70, 25 × 70, 20 × 80, 20 × 88, 26 × 88, 32 × 88, 40 × 88, and 20 × 100 (with the 88-mm apertures having semi-apochromatic triplet objectives).

Bausch & Lomb/Bushnell Performance Optics

   9200 Cody St., Overland Park, KS 66214

   Phones: 913-752-3400/800-423-3537

   Web site: www.bushnell.com

   Models: 10 × 32, 12 × 32, 16 × 32, 7 × 35, 10 × 35 (image-stabilized) 8 × 40, 7 × 42, 8 × 42, 10 × 42, 12 × 42, 8 × 43, 10 × 43, 12 × 43, 7 × 50, 10 × 50, and 8 − 15 × 50 (zoom).

Brunton Company

   620 East Monroe, Riverton, WY 82501

   Phones: 307-856-6559/800-443-4871

   Web site: www.brunton.com

   Models: 7 × 50 and 10 × 50.

Canon, Incorporated

   One Canon Plaza, Lake Success, NY 11042

   Phone: 800-652-2666

   Web site: www.canon.com

   Models: 10 × 30 and 18 × 50 (both image-stabilized).

Celestron International

   2835 Columbia St., Torrance, CA 90503

   Phone: 310-328-9560

   Web site: www.celestron.com

   Models: 7 × 35, 8 × 40, 7 × 50, 10 × 50, 8 × 56, 9 × 63, 20 × 80, and 25 × 100.

Coronado

   1674 South Research Loop, Suite 436, Tucson, AZ 85710

   Phones: 520-740-1561/866-SUNWATCH

   Web site: www.coronadofilters.com E-mail: info@coronadofilters.com

   Models: 10 × 25 and 12 × 60 (both equipped with fixed white-light solar filters).

Eagle Optics

   2120 West Greenview Dr., Suite 4, Middleton, WI 53562

   Phone: 800-289-1132

   Web site: www.eagleoptics.com

   Models: 8 × 42 and 9.5 × 44.

Edmund Scientifics

   60 Pearce Ave., Tonowanda, NY 14150

   Phone: 800-728-6999

   Web site: www.scientificsonline.com

   Model: 7 × 50 (plus many additional sizes and types by other manufacturers, the majority of which are not intended for stargazing).

Fujinon, Incorporated

    10 High Point Dr., Wayne, NJ 07470

    Phone: 719-395-8242

    Web site: www.fujinon.co.jp

    Models: 6×30, 14×40 (image-stabilized), 7×50, 10×50, 10×70, 16×70, 15×80, 25×150, and 40×150 (currently the world's largest production model binocular!).

Kowa Optimed, Incorporated

    20001 South Vermont Ave., Torrance, CA 90502

    Phone: 310-327-1913

    Web site: www.kowascope.com

    Models: 7×40 and 10×40.

Jim's Mobile, Incorporated (JMI)

    8550 West 14th Ave., Lakewood, CO 80215

    Phones: 303-233-5359/800-247-0304

    Web site: www.jimsmobile.com E-mail: info@jmitelescopes.com

    Models: Telescopic binoculars with 6-inch, 10-inch, and 16-inch apertures.

Leupold & Stephens, Incorporated

    P.O. Box 688, Beaverton, OR 97075

    Phone: 503-526-5195

    Web site: www.leupold.com92620

    Models: 8×42 and 10×50.

Meade Instruments Corporation

    6001 Oak Canyon, Irvine, CA

    Phones: 949-451-1450/800-626-3233

    Web site: www.meade.com

    Models: 7×35, 7×36, 8×40, 8×42, 10×42, 10×50, and 12×50.

Minolta Corporation

    101 Williams Dr., Ramsey, NJ 07446

    Phone: 201-825-4000

    Web site: www.minolta.com

    Models: 7×35, 8×40, 7×50, 10×50, and 12×50.

Miyauchi

    177 Kanasaki, Miano-machi, Saitama 369-1621, Japan

    Phone: 81-494-62-3371

    Models: 20×77, 20×100, and 25×141 (with apochromatic objectives).

Newcon

    3310 Prospect Ave., Cleveland, OH 44115

    Phone: 416-663-6963

    Web site: www.newcon-optik.com

    Model: 16×50 image-stabilized binocular.

Nikon

    1300 Walt Whitman Rd., Melville, NY 11747

    Phones: 631-547-4200/800-645-6687

    Web site: www.nikonusa.com

    Models: 7×35, 8×40, 10×42, 7×50, 10×50, 12×50, 10×70, 18×70, and 20×120.

Oberwerk Corporation

    2440 Wildwood, Xenia, OH 45385

    Phone: 866-244-2460

    Web site: www.bigbinoculars.com

    Models: $8/9/12/15/20 \times 60$ (choice of magnification), 11×56, 15×70, and $25/40 \times 100$ (dual magnification).

Orion Telescopes & Binoculars

    P.O. Box 1815, Santa Cruz, CA 95061

    Phones: 800-676-1343/800-447-1001

    Web site: www.oriontelescopes.com

    Models: 8×40, 8×42, 7×50, 10×50, 12×50, 8×56, 10×60, 9×63, 12×63, 15×63, 11×70, 15×70, 20×70, 11×80, 15×80, 16×80, 20×80, 30×80, and 25×100.

Parks Optical

    P.O. Box 716, Simi Valley, CA 93062

    Phone: 805-522-6722

    Web site: www.parksoptical.com

    Models: 8×42, 7×50, 10×50, 12×50, 10×52, 10×70, 11×80, 15×80, 20×80, 20×80, and 25×100.

Pentax Corporation

    35 Inverness Dr. East, Englewood, CO 80155

    Phone: 800-877-0155

    Web site: www.pentax.com

    Models: 8×40, 10×40, 7×50, 10×50, 12×50, 16×60, and 20×60.

Pro-Optic/Adorama Camera

    42 West 18th St., New York, NY 10011

    Phones: 212-741-0052/800-223-2500

    Web site: www.adorama.com E-mail: info@adorama.com

    Models: 8×42, 7×50, 10×50, 11×70, 16×70, 11×80, 20×80, 14×100, and 25×100.

Steiner/Pioneer Marketing

    97 Foster Rd., Suite 5, Moorestown, NJ 08057

    Phones: 609-854-2424/800-257-7742

    Web site: www.pioneer-research.com

    Models: 9×40, 7×50, 10×50, 8×56, 12×56, 15×80, and 20×80.

Swarovski
  2 Slater Dr., Cranston, RI 02920
  Phone: 401-734-1800
  Web site: www.swarovskioptik.com
  Models: 10 × 40 and 7 × 42.
Swift Instruments, Incorporated
  952 Dorchester Ave., Boston, MA 02125
  Phones: 617-436-2960/800-446-1116
  Web site: www.swift-optics.com
  Models: 7 × 35, 8 × 40, 7 × 42, 8 × 42, 8 × 44, 8.4 × 44, 9 × 63, 10 × 42, 7 × 50,
    10 × 50, 15 × 60, 11 × 80, and 20 × 80.
Takahashi/Land, Sea & Sky
  3110 South Shepherd, Houston, TX 77098
  Phone: 713-529-3551
  Web site: www.lsstnr.com
  Model: 22 × 60 (with apochromatic objectives).
Tasco/Bushnell Performance Optics
  9200 Cody, Overland Park, KS 66214
  Phones: 913-752-3400/800-423-3537
  Web site: www.tasco.com
  Models: 8 × 40, 8 × 42, 10 × 42, 10 × 50, 8 × 56, and 9 × 63.
Unitron, Incorporated
  170 Wilbur Place, Bohemia, NY 11716
  Phone: 631-589-6666
  Web site: www.unitronusa.com
  Models: 7 × 50, 10 × 50, 11 × 80, 20 × 80, and 25 × 100.
Vixen Optical Industries, Limited
  247 Hongo, Tokorozawa, Saitama 359, Japan
  Phone: 81-042-944-4141
  Web site: www.vixen.co.jp
  Models: 36 × 80, 20 × 125, 30 × 125, and 25 − 75 × 125 (zoom).
Zeiss (Carl), Incorporated
  1 Zeiss Dr., Thornwood, NY 10594
  Phone: 800-338-2984
  Web site: www.zeiss.com
  Models: 8 × 30, 10 × 30, 8 × 32, 10 × 32, 8 × 40, 10 × 40, 7 × 42, 8 × 42, 10 × 42,
    12 × 45, 15 × 45, 7 × 50, 8 × 56, 10 × 56, 12 × 56, and 20 × 60
    (image-stabilized).

CHAPTER NINE

Telescope Sources

Product and Contact Information for Principal Manufacturers/Suppliers

Presented here in alphabetical order is a listing of the primary sources for astronomical telescopes. Except in the case of those for some overseas manufacturers,
note that this compilation does *not* include dealers themselves, advertisements
for whom can be readily found in the various astronomical magazines such as
Sky & Telescope. Under the company's bolded name are its mailing address,
phone number/s where available, Internet web site and/or e-mail address, and a
concise listing of the telescopes offered at the time of writing. Readers should
contact them directly for copies of their latest catalogs, specifications for specific
models, current prices, shipping charges, delivery times, and warranty and return
policy. Note that most of these firms also offer a line of mounting options and
accessories in addition to the instruments themselves. In a number of cases, their
catalogs are available on-line, as well as by mail. Also listed *un-bolded* are numerous companies from the past who are either out of business entirely or are no
longer making telescopes. Many of their instruments are not only still in use by
stargazers today but are also offered from time to time on the used market. In
many cases these telescopes are considered classics, and as such are much sought
after by observers and collectors alike. No contact information is given for these
firms other than where they were or currently are located.

APM-Telescopes

Saarbrucken, Germany

Web site: www.apm-telescopes.de E-mail: apm_telescopes@web.de

Models: 3-inch to 21-inch apochromatic refractors; 16-inch to 60-inch Cassegrain and Ritchey–Chrétien reflectors; 5-inch to 24-inch Maksutov–Cassegrain catadioptrics.

Apogee, Incorporated

P.O. Box 136, Union, IL 60180

Phones: 815-568-2880/877-923-1602

Web: www.apogeeinc.com E-mail: apogeeinc@sbcglobal.net

Models: 80-mm and 90-mm RFT achromatic refractors; 80-mm apochromatic refractor; 135-mm Dobsonian reflector.

Aries Instruments Company

58 Ushakova st.kv.39, Kherson, 325026 Ukraine

Phone: 380-55-227-9653

Models: 6-inch, 7-inch, and 8-inch apochromatic refractors.

Astro-Physics, Incorporated

11250 Forest Hill Rd., Rockford, IL 61115

Phone: 815-282-1513

Web site: www.astro-physics.com

Models: 4.1-inch, 5.1-inch, and 6.1-inch "Starfire" apochromatic refractors.

Boller & Chivens Division/Perkin-Elmer Corporation

South Pasadena, CA

Models: 16-inch to 36-inch Cassegrain reflectors for advanced amateurs and professional observatories. Perkin-Elmer itself did the optics for the famed Hubble Space Telescope.

Borg Telescopes/Hutech Corporation

23505 Crenshaw Blvd., #225 Torrance, CA 90505

Phone: 310-325-5511/877-289-2647

Web site: www.sciencecenter.net/hutech

Models: 3-inch and 4-inch achromatic refractors; 3.1-inch, 3.5-inch, and 5.5-inch apochromatic refractors.

Brashear (John) Optical Company

Pittsburgh, PA

Models: achromatic refractors to 30-inches; Newtonian and Cassegrain reflectors to 72-inch aperture, for professional observatories – plus many student-observatory and amateur-class instruments (primarily refractors 4-inch to 12-inch in size). Later became J.W. Fecker (see below). The company was eventually taken over by Contraves-Goerz and operates today as Brashear LP, manufacturing custom-made, ultra-large-aperture (325-inch-class!) telescopes for astronomical research and military applications.

Bushnell Performance Optics/Bausch & Lomb

9200 Cody St., Overland Park, KS 66214

Phones: 913-752-3400/800-423-3537

Web site: www.bushnell.com

Models: 4.5-inch "Voyager" RFT reflector; 3-inch, 4.5-inch, 6-inch, and 8-inch Newtonian reflectors; 90-mm Maksutov–Cassegrain catadioptric. Also markets a series of small-aperture, low-end achromatic refractors which, unfortunately, use many plastic parts in the tube assemblies and mountings. B&L took over Criterion Manufacturing (see below) and for a number of years sold its "Dynamax" 6-inch and 8-inch Schmidt–Cassegrain catadioptrics under their name, adding a 4-inch model to the line as well.

Cave Optical Company

Long Beach, CA

Models: famed line of "Astrola" 6-inch, 8-inch, and 10-inch RFT reflectors; 6-inch, 8-inch, 10-inch, 12.5-inch, 16-inch, and 18-inch Newtonian reflectors; 8-inch, 10-inch, 12.5-inch, and 16-inch Cassegrain reflectors; 4-inch and 6-inch achromatic refractors.

Celestron International

2835 Columbia St., Torrance, CA 90503

Phone: 310-328-9560

Web site: www.celestron.com

Models: 5-inch, 6-inch, 8-inch, 9.25-inch, 11-inch, and 14-inch Schmidt–Cassegrains; 4-inch Maksutov–Cassegrain; 3.1-inch, 4-inch, and 6-inch achromatic refractors; 4-inch apochromatic refractor; 3-inch, 4.5-inch, 6-inch, 8-inch, 11-inch, 14-inch, and 17.5-inch Newtonian reflectors. In its early years, Celestron made 10-inch, 16-inch, and 22-inch Schmidt–Cassegrains, many of which are still in use today. These were followed by its famed 8-inch Schmidt–Cassegrain (known today as the Classic C8), which rapidly became the largest-selling and most popular telescope in the world after the imported 60-mm refractor.

Ceravolo Optical

Toronto, Canada

Model: 8.5-inch Maksutov–Newtonian reflector.

Cheshire Instruments

Atlanta, GA

Models: Custom-made achromatic refractors from 5-inch to 12-inch in aperture in classic-style brass tubes.

Chicago Optical

Morton Grove, IL

Models: 4-inch and 6-inch RFT reflectors.

Clark (Alvan) & Sons

Cambridgeport, MA

Models: achromatic refractors to 36-inch (Lick Observatory) and 40-inch (Yerkes Observatory) aperture, plus numerous 4-inch to 12-inch refractors for student/college observatories and amateur astronomers. Clark instruments are considered true classics and are much-sought-after collector's items in the smaller sizes today.

Coronado

1674 S. Research Loop, Suite 436, Tucson, AZ 85710.

Phones: 520-740-1561/888-SUNWATCH

Web site: www.coronadofilters.com

Models: dedicated solar telescopes – 40-mm to 140-mm aperture refractors with internal hydrogen-alpha (or calcium-K) narrow-bandpass filters.

Coulter Optical Company, Incorporated

Idyllwild, CA

Models: the world's first commercial Dobsonian reflector was Coulter's 13.1-inch Odyssey-I, followed by 8-inch, 10.1-inch, 17.5-inch, and 27-inch instruments. After Coulter went out of business, some of the Odyssey models were reintroduced for a short while by Murnaghan Instruments (see below), but the Dobsonian line was eventually discontinued.

Criterion Manufacturing Company (later Criterion Scientific Instruments)

Hartford, CT

Models: classic "Dynascope" series of 4-inch, 6-inch, and 8-inch Newtonian reflectors. The superb 6-inch RV-6 became one of the best-selling telescopes of all time and is still widely sought-after by observers and collectors today. Criterion also later manufactured 4-inch, 6-inch and 8-inch "Dynamax" Schmidt–Cassegrain catadioptrics. The company was eventually taken over by Bushnell/Bausch & Lomb (see above).

D&G Optical Company

2075 Creek Rd., Manheim, PA 17545

Phone: 717-665-2076

Web site: www.dgoptical.com

Models: 5-inch, 6-inch, 8-inch, and 10-inch achromatic refractors, plus larger sizes custom-made upon request.

DFM Engineering, Incorporated

1035 Delaware Ave., Unit D, Longmont, CO 80501

Phone: 303-772-9411

Web site: www.dfmengineering.com E-mail: sales@dfmengineering.com

Models: custom-made, mainly observatory-class Cassegrain reflectors 16-inch to 50-inch in aperture.

DGM Optics

P.O. Box 120, Westminster, MA 01473

Phone: 978-874-2985

Web site: www.erols.com/dgmoptics

Models: 4-inch to 9-inch aperture off-axis Newtonian reflectors.

Dobbins Instrument Company

Lyndhurst, OH

Models: 4.5-inch, 5-inch, 6-inch, 8-inch, and 10-inch achromatic refractors; 8-inch, 10-inch, and 12-inch Newtonian reflectors.

Discovery Telescopes

28752 Marguerite Parkway, Unit 12, Mission Viejo, CA 92692

Phone: 949-347-0142/877-523-4400

Web site: www.discovery-telescopes.com

Models: 12.5-inch, 15-inch, 17.5-inch, 20-inch, and 24-inch Dobsonian reflectors.

Ealing Optics Division/The Ealing Corporation

Cambridge, MA

Models: 12-inch to 30-inch Cassegrain reflectors.

Edmund Scientifics

60 Pearce Ave., Tonowanda, NY 14150

Phone: 800-728-6999

Web site: www.scientificsonline.com

Models: 60-mm achromatic refractor; 4.25-inch Astroscan RFT reflector (see Edmund Scientific Company below).

Edmund Scientific Company

Barrington, NJ

Models: creator of the famed 4.25-inch Astroscan RFT, which was eventually sold to Edmund Scientifics (an independent company – see above). Edmund also offered early 3-inch, 4.25-inch, 6-inch, and 8-inch basic Newtonian reflectors. In later years, an upgraded line of 3-inch, 4.25-inch, 6-inch, and 8-inch Newtonians was introduced, but then eventually discontinued when the decision was made to concentrate on industrial and military rather than hobbyist optics.

Essential Optics

Big Bear City, CA

Models: 8-inch, 10-inch, 12.5-inch, 14.25-inch, and 18-inch Newtonian reflectors.

Excelsior Optics

6341 Osprey Terrace, Coconut Creek, FL 33073

Phone: 954-574-0153

Web site: www.excelsioroptics.com

Models: 10-inch Newtonian reflectors (two).

Fecker (J.W.), Incorporated

Pittsburgh, PA

Models: 4-inch "Celestar" Newtonian reflector; 6-inch "Celestar" Maksutov-modified Cassegrain reflector. Also made many large observatory-class Newtonian and Cassegrain reflectors ranging from 24 inches to 72 inches in aperture. Successor to the John Brashear Optical Company (see above).

Fitz (Henry)

New York, NY

Models: 4-inch to 13-inch achromatic refractors by one of America's earliest telescope makers.

Galileo Telescopes

13873 SW 119th Ave., Miami, FL 33186

Phone: 800-548-537

Web site: www.galileosplace.com E-mail: customerservice@galileosplace.com

Models: 50-mm, 60-mm, 72-mm, and 80-mm achromatic refractors; 80-mm, 90-mm, 102-mm, and 120-mm Newtonian reflectors.

Goto Optical Manufacturing Company

Tokyo, Japan

Web site: www.goto.co-jp E-mail: info2@goto.co.jp

Models: 3.1-inch and 5-inch apochromatic refractors; 5-inch and 8.4-inch Newtonian reflectors; 7.9-inch coudé reflector; 17.7-inch Cassegrain reflector.

Grubb Parsons

Newcastle upon Tyne, UK

Models: 18-inch and larger Newtonian and Cassegrain reflectors for professional (and some private) observatories. This historic company had produced many world-class instruments, including the 98-inch Isaac Newton and 170-inch William Herschel telescopes.

Hardin Optical

1450 Oregon Ave., Bandon, OR 97411

541-347-4847/1-800-394-3307

Web: www.hardinoptical.com

Models: 6-inch, 8-inch, 10-inch, and 12-inch Dobsonian reflectors.

Helios Telescopes/Optical Visions Limited

Suffolk, UK

Web site: www.opticalvision.co.uk

Models: 4.5-inch, 6-inch, and 8-inch Newtonian reflectors.

Infinity Scopes, LLC

P.O. Box 69207, Tucson, AZ 85737

Phone: 520-248-0932

Web site: www.infinityscopes.com

Model: Lightweight, ultra-portable 8-inch fork-mounted Newtonian reflector.

Intes Telescopes/ITE Astronomy

16222 133rd Dr. N, Jupiter, FL 33478

Phone: 800-699-0906

Web site: www.iteastronomy.com

Models: 6-inch, 7-inch, and 9-inch Maksutov–Cassegrain catadioptrics; 6-inch and 7-inch Maksutov–Newtonian catadioptrics.

Intes Micro Company, Limited

Moscow, Russia

Phone: 7-095-126-9903

Models: 5-inch, 6-inch, 7-inch, and 8-inch Maksutov–Newtonian catadioptrics; 6-inch to 16-inch aperture Maksutov–Cassegrain catadioptrics. (Not related to Intes Telescopes.)

Jaegers (A.)

Lynbrook, NY

Models: 2-inch, 3-inch, and 5-inch achromatic refractor optical-tube assemblies. Well-known in years past for their achromatic objectives up to 6-inch in aperture by those desiring to assemble their own refractors.

Jim's Mobile, Incorporated (JMI)

8550 West 14th Ave., Lakewood, CO 80215

Phones: 303-233-5353/800-247-0304

Web site: www.jimsmobile.com E-mail: info@jmitelescopes.com

Models: 6-inch, 12.5-inch, and 18-inch "Next Generation Telescopes" Newtonian reflectors. (See also Chapter 8 for their 6-inch, 10-inch, and 16-inch binocular telescopes.)

Johnsonian Designs

3466 E. Country Rd. 20-C, Unit B20, Loveland, CO 80537

Phone: 910-219-6392

Web Site: www.johnsonian.com

Models: 8-inch, 10-inch, and 12-inch "Pop-up" Dobsonian reflectors.

Konus USA Corporation

7275 NW 87th Ave., Miami, FL 33178

Phone: 305-592-5500

Web site: www.konus.com

Models: 3.1-inch, 3.5-inch, 4-inch, 4.7-inch, and 6-inch achromatic refractors; 4.5-inch, 6-inch, and 8-inch Newtonian reflectors.

LiteBox Telescopes

1415 Kalakaua Ave., Suite 204, Honolulu, HI 96826

Phone: 808-524-2450

Web site: www.litebox-telescopes.com

Models: 12.5-inch, 15-inch, and 18-inch Dobsonian reflectors.

LOMO America, Incorporated

15 E. Palatine Rd., Unit 104, Prospect Heights, IL 60070

Phones: 847-215-8800/888-263-0356

Web site: www.lomoamerica.com E-mail: infor@lomoamerica.com

Models: 2.8-inch to 8-inch Maksutov–Cassegrain catadioptrics; 4-inch to 8-inch Maksutov–Newtonian catadioptrics – all made in Russia.

Mag 1 Instruments

16342 W. Coachlight Dr., New Berlin, WI 53151

Phone: 262-785-0926

Web site: www.mag1instruments.com

Models: 8-inch, 10-inch, 12.5-inch, and 14.5-inch "PortaBall" Newtonian reflectors.

MC Telescopes

RR #3, Box 3745, Nicholson, PA 18446

Phone: 570-942-6838

Web site: www.mctelescopes.com

Models: 10-inch, 12.5-inch, 15-inch, 16-inch, and 18-inch Dobsonian reflectors.

Meade Instruments Corporation

6001 Oak Canyon, Irvine, CA 92618

Phones: 949-451-1450/800-919-4047

Web site: www.meade.com

Models: 60-mm, 70-mm, 80-mm, 90-mm, 5-inch, and 6-inch achromatic refractors; 4-inch, 5-inch, 6-inch, and 7-inch apochromatic refractors; 3-inch to 12-inch Newtonian reflectors; 6-inch, 8-inch, and 10-inch Schmidt–Newtonian catadioptrics; 90-mm, 105-mm, 125-mm and 7-inch Maksutov–Cassegrain catadioptrics; 8-inch, 10-inch, 12-inch, 14-inch, and 16-inch Schmidt–Cassegrains; 10-, 12-, 14- and 16-inch Ritchey–Chrétien catadioptric reflectors.

Murnaghan Instruments

1781 Primrose Lane, West Palm Beach, FL 33414

Phone: 561-795-2201

Web site: www.e-scopes.cc/murnaghan_instruments_corp56469

Models: 90-mm and 4-inch achromatic refractors; 4.5-inch and 6-inch Newtonian reflectors. Also marketed Coulter Optical's Odyssey Dobsonian reflectors (see above) for a brief period, then subsequently discontinued making them.

NightSky Scopes, LLC

61432 Daspit Rd., Lacombe, LA 70445

Phone: 985-882-9269

Web site: www.nightskyscopes.com

Models: 12.5-inch, 14.5-inch, 16-inch, 18-inch, 20-inch, and 22-inch Dobsonian reflectors.

Novosibirsk Instruments/TAL Instruments

Novosibirsk, Russia

Web site: www.telescopes.co.ru

Models: 2.5-inch, 3.1-inch, 4.3-inch, 4.7-inch, and 6-inch Newtonian reflectors.

ObservatoryScope

P.O. Box 818, Ellijay, GA 30540

Phone: 706-636-1177

Web site: www.ObservatoryScope.com

Models: 20-inch to 36-inch aperture Ritchey–Chrétien reflectors.

Obsession Telescopes

P.O. Box 804, Lake Mills, WI 53551

Phone: 920-648-2328

Web site: www.obsessiontelescopes.com

Models: 12.5-inch, 15-inch, 18-inch, 20-inch, and 25-inch Dobsonian reflectors. (Also formerly made 30-inch and 32-inch aperture Dobsonians!)

Optical Craftsmen (The)

Chatsworth, CA

Models: 3-inch to 16-inch Newtonian, Cassegrain, and Newtonian–Cassegrain reflectors. Their 8-inch Discoverer Newtonian was one of the finest instruments in its aperture class ever offered to stargazers.

Optical Guidance Systems

2450 Huntingdon Pike, Huntingdon Valley, PA 19006

Phone: 215-947-5571

Web site: www.opticalguidancesystems.com

Models: 10-inch to 32-inch Cassegrain reflectors; 10-inch to 32-inch Ritchey–Chrétien reflectors.

Optical Techniques, Incorporated

Newtown, PA

Models: 4-inch and 6-inch Quantum Maksutov–Cassegrains, manufactured by former Questar (see below) employees.

Orion Optics UK

Cheshire, UK

Web site: www.orionoptics.co.uk

Models: 4.5-inch, 6-inch, 8-inch, 10-inch, and 12-inch Newtonian reflectors; 5.5-inch Maksutov–Cassegrain catadioptric. Not related to Orion Telescopes & Binoculars (next).

Orion Telescopes & Binoculars

P.O. Box 1815, Santa Cruz, CA 95061

Phones: 800-676-1343/800-447-1001

Web site: www.oriontelescopes.com

Models: 80-mm RFT achromatic refractor; 60-mm, 70-mm, 90-mm, 100-mm, and 120-mm achromatic refractors; 80-mm semi-apochromatic refractor; 100-mm apochromatic refractor; 90-mm, 102-mm, 127-mm, and 150-mm Maksutov–Cassegrain catadioptrics; 3-inch, 4.5-inch, 5.1-inch, 6-inch, 8-inch, 10-inch, and 12-inch Newtonian reflectors; 3.6-inch off-axis (unobstructed) Newtonian reflector; 4.5-inch StarBlast RFT reflector; 8-inch, 9.25-inch, and 11-inch Schmidt–Cassegrain catadioptrics (made by Celestron for Orion, under the latter's label).

Pacific Telescope Company/Synta Optical Technology Corporation (Suzhou, China)

160-11880 Hammersmith Way, Richmond, BC, Canada V7A 5C8

Phone: 604-241-7027

Web site: www.skywatchertelescope.com

Models: 3.1-inch, 3.5-inch, 4-inch, 4.7-inch, and 6-inch achromatic refractors; 80-mm, 100-mm, and 120-mm apochromatic refractors; 4.5-inch, 5.1-inch, 6-inch, and 8-inch Newtonian reflectors; 6-inch, 8-inch, and 10-inch Dobsonian reflectors; 6-inch and 7-inch Maksutov–Cassegrains.

Parallax Instruments

P.O. Box 303, Montgomery Center, VT 05471

Phone: 802-326-3140

Web site: www.parallaxinstruments.com

Models: 10-inch to 20-inch Cassegrain reflectors; 10-inch to 20-inch Ritchey–Chrétien reflectors.

Parks Optical

750 E. Easy St., Simi Valley, CA 93065

Phone: 805-522-6722

Web site: www.parksoptical.com

Models: 4.5-inch Companion RFT reflector; 6-inch, 8-inch, 10-inch, 12.5 inch, and 16-inch Newtonian reflectors; 6-inch to 16-inch Newtonian–Cassegrain reflectors.

Pentax Corporation

35 Inverness Dr. East, Englewood, CO 80155

Phone: 800-877-0155

Web site: www.pentax.com

Models: 65-mm, 85-mm, and 100-mm achromatic refractors; 65-mm, 75-mm, 85-mm, and 100-mm apochromatic refractors.

Photon Instrument, Limited

122 E. Main St., Mesa, AZ 85201

Phones: 480-835-1767/800-574-2589

Web site: www.photoninstrument.com E-mail: telescopes@photoninstrument.com

Models: 102-mm and 127-mm achromatic refractors. Photon is also one of the premier companies for the repair and restoration of classic, amateur, and professional instruments.

Questar Corporation

6204 Ingham Rd., New Hope, PA 18938

Phone: 215-862-5277/800-247-9607

Web site: www.questarcorporation.com

Models: 3.5-inch and 7-inch Questar Maksutov–Cassegrain catadioptrics. Also 12-inch custom-made Maksutov–Cassegrain observatory model upon special order. The 3.5-inch Questar was the world's first commercial catadioptric telescope, and early models are much sought-after by collectors today.

RC Optical Systems

3507 Kiltie Loop, Flagstaff, AZ 86001

Phone: 520-773-7584

Web site: www.rcopticalsystems.com

Models: 10-inch, 12.5-inch, 16-inch, and 20-inch Ritchey–Chrétien reflectors.

RVR/Asko

Rochester, NY

Models: 12.2-inch, 15.7-inch, and 17.7-inch Newtonian reflectors.

Skyscope Company, Incorporated (The)

New York, NY

Model: marvelous little 3.5-inch Newtonian reflector with 60 × eyepiece that got many stargazers of the past generation started in astronomy at a very affordable price.

Sky Valley Scopes

9215 Mero Rd., Snohomish, WA 98290

Phone: 360-794-7757

E-mail: info@skyvalleyscopes.com

Models: 12-inch to 18-inch Dobsonian reflectors.

Spacek Instrument Company

Pottstown, PA

Models: 4-inch, 6-inch, and 8-inch Newtonian reflectors; 12-inch Cassegrain reflector; 6-inch refractor; 10-inch Newtonian–Maksutov camera.

S&S Optika

Englewood, CO

Model: excellent, affordable 6-inch Newtonian reflector.

Stargazer Steve

Sudbury, Ontario, Canada

Models: beginner's 4.25-inch Newtonian reflector; 4.25-inch and 6-inch Dobsonian reflector kits.

Star Liner

Tucson, AZ

Models: 6-inch, 8-inch, and 16-inch Newtonian reflectors; 8-inch, 10-inch, 12.5-inch, and 14.25-inch Cassegrain reflectors.

StarMaster Telescopes

2160 Birch Rd., Arcadia, KS 66711

Phone: 620-638-4743

Web site: www.starmastertelescopes.com

Models: 11-inch, 12.5-inch, 14.5-inch, 16-inch, 18-inch, 20-inch, and 24-inch Dobsonian reflectors.

Starsplitter Telescopes

3228 Rikkard Dr., Thousand Oaks, CA 91362

Phone: 805-492-0489

Web site: www.starsplitter.com

Models: 8-inch, 10-inch, 12.5-inch, 15-inch, 20-inch, 24-inch, 28-inch, and 30-inch Dobsonian reflectors.

Stellarvue

11820 Kemper Rd., Auburn, CA 95603

Phone: 530-823-7796

Web site: www.stellarvue.com

Models: 3.1-inch and 4-inch achromatic refractors; 4-inch semi-apochromatic refractor; 4-inch, 6-inch apochromatic refractor.

Swift Instruments, Incorporated

952 Dorchester Ave., Boston, MA 02125

Phones: 617-436-2960/800-446-1116

Web site: www.swift-optics.com

Models: 2.4-inch and 3-inch achromatic refractors.

Takahashi/Texas Nautical Repair, Incorporated

3110 S. Shepherd Dr., Houston, TX 77098

Phone: 713-529-3551

Web site: www.takahashiamerica.com

Models: 2.4-inch, 3-inch, 3.5-inch, 4-inch, 4.2-inch, 5-inch, 6-inch, and 8-inch apochromatic refractors; 5.2-inch, 6.3-inch, and 7.9-inch Newtonian reflectors; 8.3-inch Newtonian–Cassegrain reflector; 7.1-inch, 8.3-inch, 9.8-inch, and 11.8-inch Dall–Kirkham reflectors; 10-inch Ritchey–Chrétien reflector. Also 6.3-inch, 8.3-inch, and 9.8-inch modified Newtonian short-focus reflectors with hyperbolic primary mirror and four-element corrector/field flattener lens for both wide-field photographic and visual use.

Tasco/Bushnell Performance Optics

9200 Cody St., Overland Park, KS 66214

Phones: 913-752-3400/800-423-3537

Web site: www.tasco.com

Models: 50-mm, 60-mm, and 76-mm refractors; 3-inch RFT and 4.5-inch Newtonian reflectors. Tasco was the largest and best known of numerous companies (including Jason/Empire and Mayflower) importing and marketing small achromatic refractors from the Far East. It was subsequently taken over by Bushnell.

Tectron Telescopes

Chiefland, FL

Models: 20-inch and 25-inch Dobsonian reflectors.

Telescope Engineering Company (TEC)

15730 West 6th Ave., Golden, CO 80401

Phone: 303-273-9322

Web site: www.telescopengineering.com

Models: 140-mm apochromatic refractor; 6-inch, 8-inch, and 10-inch Maksutov–Cassegrain catadioptrics; 7-inch and 8-inch Maksutov–Newtonian catadioptrics.

Teleport Telescopes

4030 N. Highway 78, Wylie, TX 75098

Phone: 972-442-5232

Web site: www.teleporttelescopes.com

Models: 7-inch and 10-inch Dobsonian reflectors.

Tele Vue Optics, Incorporated

32 Elkay Dr., Chester, NY 10918

Phone: 845-469-4551

Web site: www.televue.com

Models: 60-mm, 70-mm, 76-mm, 85-mm, 101-mm, 102-mm, and 127-mm state-of-the-art apochromatic refractors. Tele Vue also manufactures the famed "Nagler" series of highly corrected, ultra-wide-angle eyepieces designed by company founder and optical genius Al Nagler.

3-B Optical Company

Mars, PA

Models: 6-inch to 12-inch Dall–Kirkham reflector optics kits. Its ads in *Sky & Telescope* magazine caused quite a sensation with their bolded header "3 **MILES FROM MARS!**" (implying that the Red Planet could be seen that close with their optics – but actually only referring to their address here on Earth!).

Tinsley Laboratories, Incorporated

Berkeley, CA

Models: 5-inch Maksutov–Cassegrain; 12-inch Cassegrain reflector. Also manufactured large observatory-class telescopes.

TMB Optical

P.O. Box 44331, Cleveland, OH 44144

Phone: 216-524-1107

Web site: www.tmboptical.com

Models: 80-mm to 10-inch apochromatic refractors (10 different apertures).

Unitron, Incorporated

170 Wilbur Place, Bohemia, NY 11716

Phone: 631-589-6666

Web site: www.unitronusa.com

Models: 2.4-inch, 3-inch, and 4-inch achromatic refractors. Unitron refractors produced from the early 1950s to the early 1970s are considered classics, and are much sought-after today on the used market by observers and collectors alike. At one time Unitron offered as many as 26 models, ranging from 1.6-inch to 6-inch in aperture!

Unertl (John) Optical

Pittsburgh. PA

Models: 60-mm to 90-mm short-focus achromatic refractors (essentially spotting scopes); custom-made reflectors of various sizes and types, including an 8-inch aperture modified Cassegrain reflector with stationary eyepiece.

VERNONscope/Brandon Optical

Ithaca, NY

Models: 94-mm and 130-mm apochromatic refractors. Brandon made an excellent 3-inch achromatic refractor in the past and was famous for its orthoscopic eyepieces, which are still manufactured by VERNONscope today.

Vixen North America/Tele Vue Optics

32 Elkay Dr., Chester, NY 10918

Phone: 845-469-8660

Web site: www.vixenamerica.com

Models: 60-mm, 80-mm, 90-mm, 102-mm, 4.7-inch, and 5.1-inch achromatic refractors; 80-mm, 4-inch, 4.5-inch, and 5.1-inch apochromatic refractors; 8-inch Cassegrain reflector.

William Optics USA

4200 Avenida Sevilla, Cypress, CA 90630

Phone: 714-209-0388

Web site: www.william-optics.com E-mail: wo@william-optics.com

Models: 66-mm, 80-mm, and 105-mm apochromatic refractors.

Zeiss (Carl) Jena

Jena, Germany

Web site: www.zeiss.de

Zeiss withdrew from the amateur telescope market in 1994 and discontinued making instruments smaller than 16 inches in aperture in order to concentrate on their world-class large-observatory models. Superb Zeiss apochromatic refractors 4-inch (100-mm), 5-inch (130-mm) and 6-inch (150-mm) in aperture, plus a 7-inch (180-mm) Maksutov–Cassegrain, were marketed briefly through their US distributor, Seiler Instrument & Manufacturing Company in St. Louis, MO. Many of these instruments are actively in use today by observers who wanted to have the finest optics money could buy – from a truly world-class optical firm.

Part II

USING ASTRONOMICAL TELESCOPES AND BINOCULARS

CHAPTER TEN

Observing
Techniques

Training the Eye

It has often been said that the person behind the eyepiece of a telescope or pair of binoculars is far more important than the size or type or quality of the instrument itself. An inexperienced observer may look at the planet Jupiter and perhaps detect its two major dark equatorial bands, while an experienced one will typically see more than a dozen belts and bands using the very same telescope at the same magnification. Again, a novice may glimpse a nebula as a barely visible ghostly glow in the eyepiece, while a seasoned observer will see intricate details and even in some cases various hues. It's all a matter of training the eye – and along with it the brain that processes the images formed by it from the telescope itself.

Sir William Herschel – the greatest visual observer who ever lived – long ago advised: "You must not expect to *see at sight*. Seeing is in some respects an art which must be learned." He also pointed out that "When an object is once discovered by a superior power, an inferior one will suffice to see it afterwards." (In other words, the eye–brain combination has been alerted to its presence.) And another great observer from the past, William Henry Smyth, stated that "Many things, deemed invisible to secondary instruments, are plain enough to one who 'knows how to see them'."

There are four distinct areas in which an observer's eye can be trained to see more at the eyepiece (whether of a telescope or of a pair of binoculars). Let's start with *visual acuity* – the ability to see or resolve fine detail in an image. There's no question that the more time you spend at the eyepiece, the more such detail you will eventually see! Even without any purposeful training plan in mind, the eye–brain combination will learn to search for and see ever-finer detail in what it is viewing. But this process can be considerably speeded up by a simple exercise repeated daily for a period of at least several weeks. On a piece of white paper, draw a circle about 3 inches in diameter. Then, using a soft pencil, randomly draw various markings within the circle, ranging from broad patchy shadings to fine lines and points. Now place the paper at the opposite side of a room at a distance of at least 20 feet or so, and begin drawing what you see using the unaided eye. Initially, only the larger markings will be visible to you – but as you repeat this process over a period of time, you'll be able to see more and more of them!

Taking this a step further, cut out that white disk and attach it onto a black background. Next, darken the room and illuminate the image with a broad-beam, low-intensity flashlight aimed at it. Doing this more closely simulates the view of a planetary disk seen against the night sky through a telescope. Tests have shown improvements in overall visual acuity by a factor of as much as 10 using these procedures! Not only will you see more detail on the Sun, Moon, and planets as a result of this practice, but you will also be able to resolve much closer double stars than you were previously able to.

A second area of training the eye–brain combination involves the technique of using *averted (or side) vision* in viewing faint celestial objects. This makes use of the well-known fact that the outer portion of the retina of the eye contains receptors called *rods*, and that these are much more sensitive to low levels of illumination than is the center of the eye, which contains receptors known as *cones*. (See the discussion below involving color perception by the cones.) This explains the common experience of driving at night and seeing objects out of the corner of the eye appearing brighter than they actually are if you turn and look directly at them.

Applied to astronomical observing, averted vision is used in detecting faint companions to double stars, and dim stars in open and globular clusters. But it's especially useful (and most obvious) in viewing low-surface-brightness targets such as nebulae and galaxies, where increases in apparent brightness of from two to two-and-a-half times (or an entire magnitude) have been reported! Once having centered such an object in the field of view, look to one side of it (above or below also works), and you'll see it magically increase in visibility. (Be aware that there is a small dark void or "dead spot" in the fovea between the eye and ear that you may encounter in going that direction.)

One of the most dramatic examples of the affect of averted vision involves the so-called "Blinking Planetary" – a name I coined many years ago in a *Sky & Telescope* magazine article about it. Also known as NGC 6826, it's located in the constellation Cygnus and is easily visible in a 3- or 4-inch glass. Here we find an obvious bluish-green tenth-magnitude nebulosity some 27 seconds of arc in size surrounding a ninth-magnitude star. Staring directly at the star, there's no sign of the nebulosity itself. On switching to averted vision, the nebulosity instantly appears and is so bright that it drowns out the central star. Alternating back and forth between direct and averted vision results in an amazing apparent blinking-on-and-off effect!

A third important area involving the eye–brain combination is that of *color perception*. At first glance, all stars look white to the eye. But upon closer inspection, differences in tint among the brighter ones reveal themselves. The lovely contrasting hues of ruddy-orange Betelgeuse and blue-white Rigel in the constellation Orion is one striking example in the winter sky. Another can be found in the spring and summer sky by comparing blue-white Vega in Lyra and orange Arcturus in Bootes. Indeed, the sky is alive with color once the eye has been trained to see it! Star color, by the way, is primarily a visual indication of surface temperature: ruddy ones are relatively cool while bluish ones are quite hot. Yellow and orange suns fall in between.

While the rods in the edge of the eye are light-sensitive, they are essentially colorblind. Thus, for viewing the tints of stars (whether single, double, or multiple) and other celestial wonders, direct vision is employed – making use of the color-sensitive cones at the center of the eye. Stare directly at an object to perceive its color and off to the side to see it become brighter (unless it's already a bright target such as a planet or naked-eye star). It should be mentioned here that there's a peculiar phenomenon known as the Purkinje Effect that results from staring at red stars – they appear to increase in brightness the longer you watch them!

One final area involving preparing the eye to see better is that of *dark adaptation*. It's an obvious fact that the eyes needs time to adjust to the dark after coming out of a brightly lit room. Two factors are at play here. One is the dilation of the pupils themselves, which begins immediately upon entering the dark and

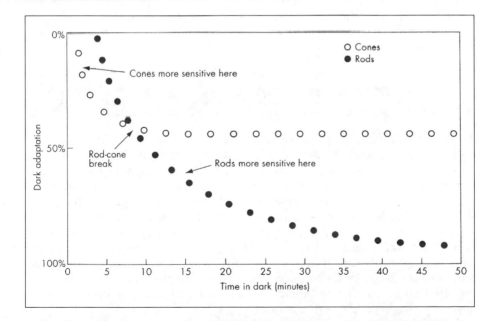

Figure 10.1. Dark-adaptation times for both the color-sensitive cones (open circles) at the center of the eye and the light-sensitive rods (black dots) around the outer part of the eye. At the "rod–cone break" some 10 minutes after being in the dark, the sensitivity of the cones levels off and remains unchanged. The rods, however, continue to increase their sensitivity to low light levels, with complete dark adaptation taking at least four hours! For all practical purposes, the eye is essentially dark-adapted in about 30 to 40 minutes.

continues for several minutes. The other involves the actual chemistry of the eye, as the hormone rhodopsin (often called "visual purple") stimulates the sensitivity of the rods to low levels of illumination. The combined result is that night vision improves noticeably for perhaps half an hour or so (and then continues to do so very slowly for many hours following this initial period). This is why the sky looks black on first going outside, but later looks gray as you fully adjust to the dark. In the first instance, it's a contrast effect; and in the second, the eye has become sensitive to stray light, light pollution, and the natural airglow of the night sky itself that were not seen initially.

Stargazers typically begin their observing sessions by viewing bright objects such as the Moon and planets first and moving to fainter ones afterward, giving the eye time to dark-adapt gradually and naturally. This is mainly of value in observing the dimmer deep-sky objects such as nebulae and galaxies. Stars themselves are generally bright enough to be seen to advantage almost immediately upon looking into the telescope. Exceptions are faint pairs of stars and dim companions to brighter stars (where the radiance of the primary often destroys the effect of dark-adaptation). White light causes the eye to lose its dark-adaptation but red light preserves it, making it standard practice to use red illumination to read star charts and write notes at the eyepiece. Another helpful procedure is to wear sunglasses (preferably polarized) when venturing outside on a sunny day if you plan to look for "faint fuzzies" that evening. It's been shown that bright sunlight – especially that encountered on an ocean beach or near other reflecting bodies of water and on snow – can retard the eye's dark adaptation for as long as several days!

Sky Conditions

A number of atmospheric and related factors affect the visibility of celestial objects at the telescope. In the case of the Moon, planets, and double stars, the most important of these is atmospheric turbulence or *seeing*, which is an indication of the steadiness of the image. On some nights, the air is so unsteady (or "boiling", as it's sometimes referred to) that star images appear as big puffy, shimmering balls, and detail on the Moon and planets is all but non-existent. This typically happens on nights of high atmospheric *transparency* – those having crystal-clear skies, with the air overhead in a state of rapid motion and agitation. On other nights, star images are nearly pinpoints with virtually no motion, and fine detail stands out on the Moon and planets like an artist's etching. Such nights are often hazy and/or muggy, indicating that stagnant, tranquil air lies over the observer's head and that seeing is superb.

One of the most dramatic and revealing accounts of the changing effects of seeing-conditions upon celestial objects comes from the great double-star observer S.W. Burnham, in the following account of the famed binary system Sirius (α Canis Majoris): "An object glass of 6-inch one night will show the companion to Sirius perfectly: on the next night, just as good in every respect, so far as one can tell with the unaided eye, the largest telescope in the world will show no more trace of the small star than if it had been blotted out of existence."

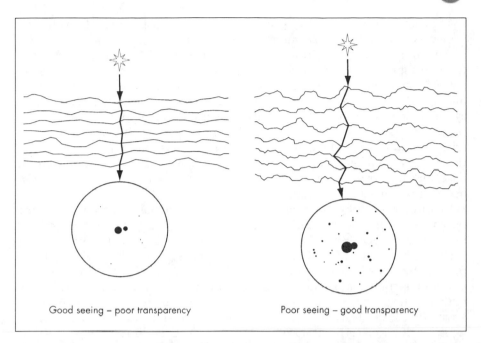

Good seeing – poor transparency   Poor seeing – good transparency

Figure 10.2. On nights of good "seeing" or atmospheric steadiness, the air above the observer is tranquil and lies in relatively smooth layers, often resulting in somewhat hazy skies. This allows starlight to pass through undisturbed and produces sharp images at the eyepiece. In poor seeing, the atmosphere is very turbulent, typically bringing with it crystal-clear skies and an ideal time to view faint objects. But the resulting star images are often blurry, shimmering balls of light, making such nights largely useless for seeing fine detail on the Moon and planets or for splitting close double stars.

Various *seeing scales* are employed by observers to quantify the state of atmospheric steadiness. One of the most common of these uses a 1-to-5 numerical scale, with 1 indicating hopelessly turbulent blurred images and 5 stationary razor-sharp ones. The number 3 denotes average conditions. Others prefer a 1-to-10 system, with 1 again representing very poor and 10 virtually perfect seeing, respectively. (In some schemes, the numerical sequence is reversed, with lower numbers indicating better and higher numbers poorer seeing.) While casual stargazing can often be done even in less than average-quality seeing, many types of observing such as sketching or imaging fine lunar and planetary detail, or splitting close double stars, requires good to excellent conditions.

Another factor affecting telescopic image quality is that known as "local seeing" – or the thermal conditions in and around the telescope itself. Heat radiating from driveways, walks and streets, houses and other structures (especially on nights following hot days) plays a significant role in destroying image quality that's totally unrelated to the state of the atmosphere itself. For this reason, observing from fields or grassy areas away from buildings and highways gives the best chance of minimizing local seeing.

The cooling of the telescope optics and tube assembly is especially critical for achieving sharp images. Depending on the season of the year, it may take up to an hour or more for the optics (especially the primary mirror in larger reflectors) to reach equilibrium with the cooling night air. During this cool-down process, air currents within the telescope tube itself can play absolute havoc with image quality, no matter how good the atmospheric seeing is – even in closed-tube systems such as the popular Schmidt–Cassegrain catadioptric. Reflecting telescopes should have tubes at least several inches larger than the primary mirror itself to allow room for thermal currents to rise along the inside of the tube rather than across the light path itself. Surprisingly, even the heat radiating from the observer's body can be a concern here, particularly with Dobsonian reflectors having open-tubed truss designs.

No discussion of sky conditions and their impact on observing would be complete without mentioning the hindering effects of bright lights – both natural and man-made. Especially around the time of full Moon, bright moonlight not only destroys the observer's dark adaptation (discussed above) but also wipes out many of the sky's faint wonders such as nebulae and galaxies. A modern accompaniment is the menace of light pollution – illumination from ever-more homes, office buildings, shopping malls, and cars lots, directed skyward instead of downward where it's actually needed. This has much the same result as bright moonlight, as it illuminates the atmosphere through which the observer must look. But additionally in the case of artificial lighting, haze and passing clouds intensify its impact by bouncing it back down into telescopes, binoculars, and observers' eyes. Fortunately, the planets and brighter stars (including variables and doubles), as well as the Moon itself, are largely immune to all this, and therefore bright nights are not necessarily a total loss for observing. Readers interested in learning more about the subject of light-pollution and ways to combat it should contact the US-based International Dark-Sky Association (IDSA) at www.darksky.org or by e-mail at ida@darksky.org.

Record Keeping

The annals of both amateur and professional astronomy attest to the personal as well as scientific value of keeping records of our nightly vigils beneath the stars. From the former perspective, an account of what has been seen each night can bring pleasant memories as we look back over the years at our first views of this or that celestial wonder – or when we shared their very first look at the Moon or Jupiter or Saturn with loved ones, friends, and even total strangers. Our eyepiece impressions written and/or sketched on paper, or perhaps recorded on audio tape and/or electronically imaged, can provide many hours of nostalgic enjoyment in months and years to come.

From a scientific perspective, you may become involved in searching for comets or patrolling the sky for novae, or monitoring the brighter spiral galaxies for possible supernova outbursts. Even negative observations may be of value to professional astronomers. Often has the call gone out to the astronomical community in the various magazines, journals, and electronic media asking if anyone happened to be looking at a certain object or part of the sky on a given date and at a particular time. If you happened to be at "the right place at the right time" indicated but noted nothing unusual in your observing log, that is still a fact of real importance to researchers. (This frequently happens in attempting to determine when a nova in our galaxy or a supernova in a neighboring galaxy first erupted.) And, of course, there's always the possibility that you will be the first to see and report something new in the sky yourself!

The information in your logbook should include the following: the date, and beginning and end times of your observing session (preferably given in Universal Time/Date); telescope size, type, and make used; magnification/s employed; sky conditions (seeing and transparency on a 1-to-5 or 1-to-10 scale, along with notes on passing clouds, haze, moonlight, and other sources of light-pollution); and finally a notation of each object seen (accompanied, if you're so inclined, by sketches, photographs, and/or electronic images). And here, an important point

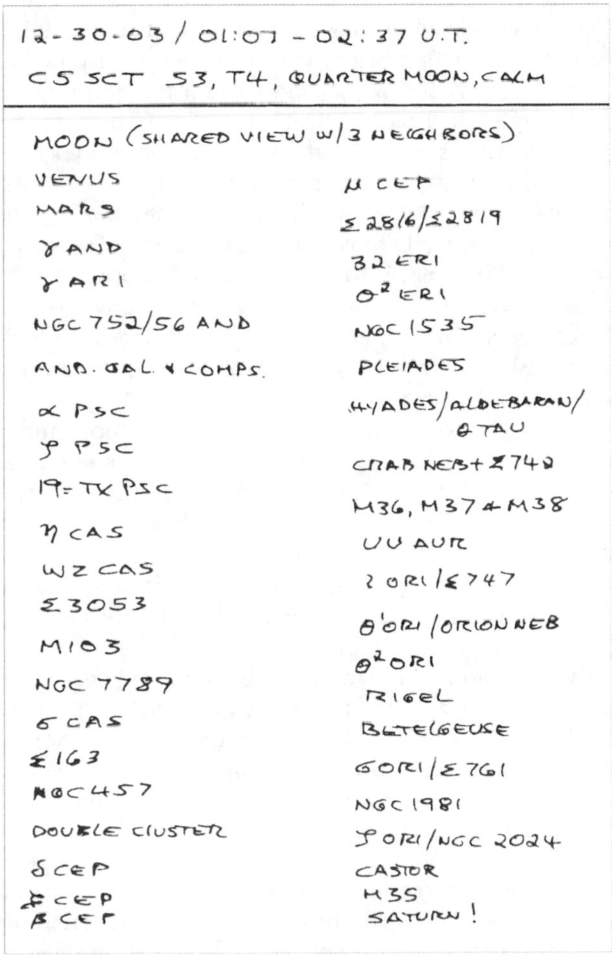

Figure 10.3. A busy night's entry from the author's personal observing logbook. The date and times are given in Universal Time (U.T.) – that of the Greenwich, England, time zone. A 5-inch Celestron Schmidt–Cassegrain catadioptric telescope (C5 SCT) was used under conditions of average seeing (S) and good transparency (T), and the sky was brightened by the light of a first-quarter Moon. All the targets viewed on this particular night are celestial showpieces! Normally, more time should be given to viewing fewer objects than shown here, in order to fully enjoy and appreciate the cosmic pageantry.

```
12-30-03 / 01:07 - 02:37 U.T.
C5 SCT  S3, T4, QUARTER MOON, CALM

MOON (SHARED VIEW W/3 NEIGHBORS)
VENUS                    μ CEP
MARS                     Σ2816/Σ2819
γ AND                    32 ERI
γ ARI                    O² ERI
NGC 752/56 AND           NGC 1535
AND. GAL. & COMPS.       PLEIADES
α PSC                    HYADES/ALDEBARAN/
ψ PSC                           θ TAU
19=TX PSC                CRAB NEB+Σ742
η CAS                    M36, M37 & M38
WZ CAS                   UU AUR
Σ3053                    ? ORI/Σ747
M103                     θ¹ORI /ORION NEB
NGC 7789                 θ² ORI
ε CAS                    RIGEL
Σ163                     BETELGEUSE
NGC 457                  σ ORI/Σ761
DOUBLE CLUSTER           NGC 1981
δ CEP                    ζ ORI/NGC 2024
ε CEP                    CASTOR
β CEP                    M35
                         SATURN !
```

should be borne in mind regarding record-keeping at the telescope: limit to an absolute minimum the amount of time you spend logging your observations (using a red light to preserve your dark adaptation when you do). Some observers spend far more time writing about what they see at the eyepiece than they actually do viewing it!

Observing Sites and Observatories

Where you use your telescope (and less so your binoculars) is a very important consideration. Some observers have no choice in this matter, being confined for various reasons to viewing from a balcony or a driveway or a rooftop, or even through an opened upstairs bedroom window – all of which are considered "taboo"! We've already mentioned that images can be badly affected by heat rising from paved surfaces and radiating from buildings. Probably worst of all is observing from a rooftop, as some observers living in apartments in large cities do to get above the lights and surrounding buildings. After a warm day, a tarred roof surface will radiate heat long into the night, engulfing both observer and telescope in a sea of unstable air. And while looking through open windows – especially during the winter months when it's cold outside and warm inside – has always been considered useless because of the temperature gradient, some observers (typically those using long-focus refractors) have been able to get reasonably sharp images by sticking the telescope tube as far outside the window as possible. Needless to say, sky visibility in any case is quite limited! And as for balconies, any movement on or anywhere around it (even someone walking inside the house or apartment) will typically result in annoying gyrations of the images seen in the telescope.

With light pollution ever on the increase, more and more stargazers are traveling with their telescopes to find dark skies. In some cases, their destinations are sites designated for astronomy clubs; in others they are city, county, state, or national parks. While this has its advantages – particularly for observers whose main interest is viewing faint deep-sky wonders – the inconvenience of packing up the telescope, accessories, and necessities such as water and proper clothing, and then driving there and back, limit the number of observing sessions compared with simply stepping out into a backyard. And for those who believe that only the Moon and planets can be observed from urban areas, I have on many occasions viewed – and also shown to others – some of the brighter galaxies such as that in Andromeda (M31) from the heart of several major cities. Two helpful references here are *Visual Astronomy in the Suburbs* by Anthony Cooke (Springer, 2003) and *Urban Astronomy* by Denis Berthier and Klaus Brasch (Cambridge University Press, 2003).

Fortunate indeed is the observer who has a proper shelter for his or her telescope – one where the instrument can be left safely outdoors, protected from the elements, and be ready to use almost immediately upon demand. The *domed observatory* is the best-known and most aesthetic-looking of such structures, and

Figure 10.4. A charming example of the classical domed observatory. The most aesthetic-looking of the various types of telescope shelters, it offers the greatest protection from wind and stray light. However, it also has the most limited view of the sky (owing to its narrow slit) and is the most costly of all the types of housing. Photo by Sharon Mullaney.

offers the maximum protection from wind and stray lights. But of the various types, it's also the most costly, requires the longest cool-down time (for the temperature inside the dome to reach equilibrium with that outside), and provides only a limited vista of the sky through its narrow slit.

An alternative to the dome is the *roll-off roof observatory*, in which the entire top of the structure rolls back on tracks to reveal the whole visible heavens. Depending on the height of its walls, it may offer only limited protection from wind and lights. A delightful compromise between these two types is the *flip-top roof observatory*. Here, a structure with low walls and hinged peak roof splits into two sections – one half typically swinging to the east, the other to the west. Chains or ropes control how high or low the halves extend, the observer adjusting them to reveal whatever part of the sky is being viewed, while at the same time using them as a shield against wind and lights.

Mention should also be made of a unique telescope shelter first built by famed stargazer Leslie Peltier to house his short-focus 6-inch refractor. Referred to as the *merry-go-round observatory*, the observer rides sitting in cushioned comfort in a chair with both the instrument and the structure housing it as they circle in

a

b

Figure 10.5a & b. A typical roll-off roof observatory – this one housing a classic Alvan Clark 8-inch refractor, shown with the roof in both its opened (top) and closed (below) position. This structure provides the maximum coverage of the sky, but both the observer and telescope are exposed to the wind and elements (to a degree that depends on how high the walls are) and there's little protection against stray light. Photos by Sharon Mullaney.

azimuth on a track to reveal a particular part of the sky. The telescope itself points skyward through a kind of trap door on the flat roof of the box-shaped observatory, and is mounted within in such a way that very little head motion is required to look into the eyepiece, no matter where it's pointed. A number of other observers have constructed their own versions of Peltier's design, and they all consider this the ultimate in a personal telescope shelter. For more about the merry-go-round observatory, see Peltier's delightful autobiography *Starlight Nights* (Sky Publishing, 2000).

It should be pointed out here that many of the classic observers of the past – including the greatest observational astronomer of them all, Sir William Herschel – worked in the open night air unprotected by any kind of structure. So, too, do a majority of stargazers today, including the author, who has extensively used all three types of observatories (and on one occasion Peltier's merry-go-round creation itself!) over the years as both an amateur and professional astronomer.

Personal Matters

There are a number of little-recognized factors that impact the overall success of an observing session at the telescope. One concerns proper dress. This is of particular importance in the cold winter months of the year, when observers often experience sub-zero temperatures at night. It's impossible to be effective at the eyepiece – or even just to enjoy the views – when you're numb and half-frozen to death! Proper protection of the head, hands, and feet are especially critical during such times, and for thermal insulation of the body in general several layers of clothing are recommended as opposed to one heavy one. During the summer months, the opposite problem arises, as observers attempt to stay cool. In addition to having very short nights at this time of the year, there's the added annoyance of flying insects together with optics-fogging humidity and dew. (See the discussion on dew caps and heated eyepieces in Chapter 7.)

Another concern is proper posture at the telescope. It's been repeatedly shown that observers see more from a comfortably seated position than when standing, twisting, or bending at the eyepiece! If you must stand, be sure that the eyepiece/focuser is at a position where you don't have to turn and strain your neck, head, or back to look into it. This is especially important in using large reflectors, which often require a ladder just to reach the eyepiece! And, while not as critical, the same goes for positioning finder scopes where they can be used without undue contortions.

Proper rest and diet both play a role in producing a pleasurable observing session. Attempting to stargaze when you're physically and/or mentally exhausted is guaranteed to leave you not only frustrated, but also looking for a buyer for your prized telescope! Even a brief cat-nap before going out to observe after a hectic day is a big help here. Heavy dinners can leave you feeling sluggish and unable to function alertly at the telescope. It's much better to eat after you're done stargazing – especially so since most observers find themselves famished then (particularly on cold nights!). Various liquid refreshments such as tea, coffee, and hot chocolate can provide a needed energy boost (and warmth when desired).

And while alcoholic drinks such as wine do dilate the pupils, technically letting in more light, they also adversely affect the chemistry of the eye. This reduces its ability to see fine detail on the Moon and planets or resolve close double stars, and especially its sensitivity in viewing "faint fuzzies" such as dim nebulae and galaxies!

Finally, there's the very important matter of *preparation*. This is not just being aware of what objects are visible on a given night at a particular time of year, or of which of them can be seen from your site and with your instrument, or deciding on those you plan to observe this time out. It goes beyond this to understanding something about the physical nature of the wonder you're looking at – facts such as its type, distance, physical size, luminosity, mass, temperature, velocity towards or away from you, speed of rotation, composition, age, and place in the grand cosmic scheme of things. In other words, as stargazers we must "see" with our minds as well as our sight. (To better grasp the importance and purpose of preparation before going to the eyepiece, see the beautiful mandate found in the opening lines of Chapter 12, as set forth by the classic observer Charles Edward Barns.)

CHAPTER ELEVEN

Solar System Observing

In this chapter we review various aspects of observing objects within our Solar System using telescopes and binoculars – both for pleasure and also for making useful observations. So vast is this field that numerous entire books have been written about each and every one of the various objects discussed. Because of space limitations, the coverage given here, of necessity, provides only an overview of this subject. References for further study and contact information for various organizations in this field are also given.

Sun

We begin with our dazzlingly bright *daytime star*, the Sun, which lies at an average distance from us of 93,000,000 miles (or 8 light-minutes). Now here's one celestial object that supplies far too much light rather than not enough (which is the usual complaint of stargazers!) – so much so that it's actually a very dangerous object to observe without proper precautions. Serious eye damage or even permanent blindness can result from attempting to look at our star directly, or from using inadequate filtering techniques. While solar filters were briefly discussed in Chapter 7, it needs to be emphasized here that the *finders* of telescopes must also be capped whenever you are viewing the Sun, as they too can do damage to your eye, hair, skin, and even clothing. And binoculars, of course, must have proper solar filters secured over both of their objective lenses (again, as with telescopes, *not* over their eyepieces!) for safe viewing.

Figure 11.1. A full-aperture optical-glass solar filter mounted over the objective of an 80-mm refractor. (Note the capped finder – a very vital precaution!) While more expensive than Mylar® solar filters, these provide better image contrast and a more natural-looking yellowish-orange Sun rather than a blue one. Securely mounted, such full-aperture devices are completely safe for viewing our Daytime Star – in contrast to the highly dangerous practice of placing a filter over the eyepiece itself. Courtesy of Orion Telescopes & Binoculars.

The most obvious visual features of the Sun even in a 2- or 3-inch glass at 25× to 30× are the *sunspots*, with their dark inner *umbra* and lighter outer *penumbra*. (A few of the very largest ones are visible to the unaided eye on occasion and many are visible in binoculars.) Not only do they steadily change in size and shape, but they continually parade across the face of the Sun from day to day as it rotates. Near the limb, they become obviously foreshortened and some even show an apparent depression in the *photosphere* (the visible surface) at their centers. As is well known, sunspots come and go in a period that averages 11 years – the so-called *sunspot cycle*. Near maximum, the Sun's face may be littered with them, often in large groups, while at minimum it may be difficult to find any at all.

On rare occasions, a brilliant *white-light flare* will suddenly make its appearance within a sunspot, cutting across it in minutes as it rises up into the Sun's outer atmosphere. Other things to look for are the *limb darkening* towards the edge of the Sun, giving it a very striking three-dimensional shape and looking like the vast sphere it actually is. There are also the minute *granulations* – tiny thermal convective cells in the photosphere that are continually churning, appearing and disappearing, and that are most conspicuous in the areas of limb darkening. A 4- or 5-inch glass at 50× or more will readily show them when seeing

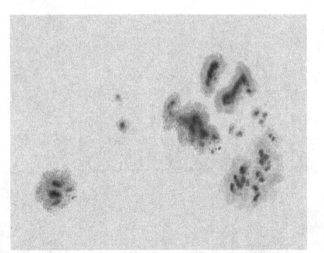

Figure 11.2. A highly-detailed, stylized sunspot sketch based on a famous classical observation by Samuel Pierpont Langley in 1870 using a 13-inch refractor at very high magnification under superb seeing conditions. Note the dark inner umbras, and the lighter wispy outer penumbras. The "freckles" surrounding the sunspots itself represent solar granulation, which are convective cells in the Sun's photosphere (or visible surface).

conditions are good. Somewhat similar are the *faculae*, which are hotter and therefore brighter patches in the photosphere that are typically seen near sunspots and, as with granulation, are most obvious near the darkened limb.

All of the above are features of the Sun as seen in visible or "white" light. With the advent of affordable hydrogen-alpha filters (and even designated solar telescopes with such filters built into them – see Chapter 4), it's become possible for amateur astronomers to look deeper into our star's seething layers and also to view its huge arching *prominences*. The latter can not only be seen rising majestically from the edge of the Sun, but also moving across its face as giant, snake-like, dark ribbons. If you're looking for a celestial object that "does something" while you watch it, there's none more dynamic or spectacular than the Sun viewed in hydrogen-alpha light!

On rare occasions, *transits* of the Sun by the inner planets Mercury and Venus can be seen. Of the two, those of Mercury are the more common; the last one was on May 7, 2003, while the next will occur on November 8, 2006. At such times, the tiny planet appears as a black dot about 10 seconds of arc in size (requiring high-powered binoculars or a small telescope to be seen) slowly moving across the Sun over a period of several hours. Transits of Venus are incredibly rare, and when they do happen they take place in pairs eight years apart. The last set occurred in 1874 and 1882, while the first of the current pair was widely observed on June 8, 2004. At that time, Venus looked like a huge round black sunspot and was easily visible even in binoculars. The second transit of this set is scheduled

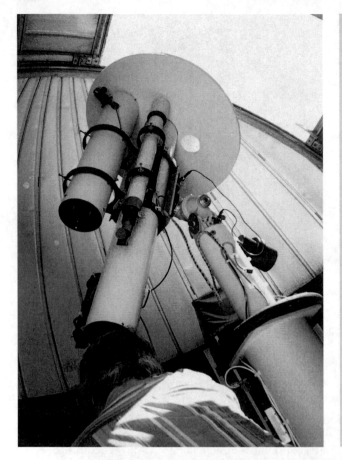

Figure 11.3. The author at the working end of a 6-inch refractor equipped with a hydrogen-alpha solar filter. Note the circular disk at the top of the tube (through which the filtered objective end of the telescope itself protrudes) covering both the finder and an auxiliary short-focus reflector (a very important precaution!). This also serves to block direct sunlight from falling on the observer. Photo by Sharon Mullaney.

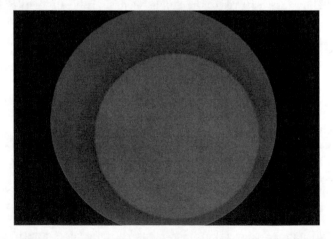

Figure 11.4. A view of the Sun made by simply pointing a camera through the eyepiece of the telescope in Figure 11.3. Careful inspection of the Sun's off-center image reveals several tiny prominences along the limb, especially in the 9- to 10-o'clock position. Visually they are much more prominent than shown here, and their graceful movement can actually be seen by patient watching. Photo by Sharon Mullaney.

for June 5, 2012. It should also be mentioned here that transits of dark objects other than Mercury and Venus themselves across the face of the Sun have been reported ever since the invention of the telescope (and there are even some pre-telescopic naked-eye accounts). Although more than 600 have appeared in the astronomical literature over the past three centuries, their nature is still a mystery. So observers should always be on the lookout for such anomalous sightings. Note that balloons and artificial satellites drift quickly over the face of the Sun, while the dark objects that have been seen move slowly, much like the planets themselves. And here again, of course, a proper solar filter is required when observing transits, whatever their source!

One last area of interest concerning the Sun is that of *solar eclipses*, in which the Earth passes through the Moon's narrow shadow-cone. These may be partial, total, or annular (with a ring of light still surrounding the Moon at *totality*) in nature. It's the total solar eclipse that's without question nature's grandest spectacle, and which brings with it so many exciting events, all happening within the precious few minutes of totality. These include the sudden visibility of the Sun's delicate but magnificent *corona*, scarlet prominences dancing along its limb, and the well-known "diamond-ring" effect. There's also the dramatic turning of daylight into darkness, the planets and brighter stars making their appearance, and a sudden drop in air temperature as the Sun's light is extinguished. Except within the narrow interval of totality itself, solar filters are absolutely essential for safely viewing an eclipse – whether with the unaided eye, with binoculars (which typically provide the best view), or through a telescope itself.

Moon

At an average distance from us of 239,000 miles, our lovely satellite is the nearest of all celestial bodies (with the exception of "Earth-grazing" asteroids!). And as such, it offers so many varied, fantastic things to see – even in a pair of binoculars. Aside from the obvious dark lunar "seas" (or *maria*) and bright lunar highland areas, there are majestic mountain ranges, vast walled plains, craters, craterlets, pits, domes, rilles, clefts, faults, and rays. As the Moon advances through its monthly *phases*, the interplay of light and shadow across this alien landscape is ever-changing, not only from hour to hour but from minute to minute. Many observers assume that when a given phase, such as first quarter (or half-full), repeats itself each month, its features will look exactly the same. Yet, owing to the subtleties of orbital dynamics, the Moon does not present the exact same appearance for another seven years! With the exception of the bright lunar rays, which are most prominent when the Moon is full, surface features are best seen along the ever-advancing *terminator* – the dividing line between night and day. Here, sunlight coming in at a very low, glancing angle casts long, dramatic shadows across the landscape, greatly exaggerating vertical relief. The Moon is perhaps most striking when around first or last quarter.

One of the most fascinating – and controversial – aspects of viewing the Moon is the subject of *transient lunar phenomena* or TLPs. Ever since the invention of

Figure 11.5. The Moon seen near its half-full phase, imaged through an 11-inch Schmidt–Cassegrain catadioptric. A few days on either side of the quarters is the best time to view surface features such as craters, mountains, and valleys. It's then that sunlight comes in at a glancing angle, casting long dramatic shadows that exaggerate vertical relief (especially along the terminator, or dividing line between day and night on the Moon). Here's a fascinating alien world right up close, to explore with binoculars and telescopes of all sizes! Courtesy of Dennis di Cicco.

the telescope, both amateur and professional astronomers have reported seeing flashes, obscurations, colorations – and even moving lights on its surface! These have been attributed to everything from an over-active imagination to atmospheric refraction, outgassing of sub-surface water, volcanic activity, and meteorite impacts. A classic example (which I have witnessed myself) involving the crater Plato is the slow disappearance of craterlets from one end of its dark floor across to the other side, and then their equally slow reappearance in reverse order – all over a period of just a few hours. Another apparently active area is the crater Aristarchus, where reddish glows have frequently been seen – by everyone from Sir William Herschel to orbiting Apollo astronauts. The key word here is "watchfulness" whenever you're observing the Moon!

As our satellite moves continuously eastward around the sky in its orbit (by roughly its own diameter each hour), it occasionally passes in front of planets, asteroids, stars, and various deep-sky objects such as star clusters, resulting in an *occultation*. The disappearance and reappearance of these objects can be quite dramatic – especially for the planets, where the rings of Saturn or the satellites of Jupiter can be seen being gobbled up and swallowed by this huge orb. In the case of stars, they instantly snap off and then later back on, like someone flipping a light switch. (For some close double stars, this will happen in steps, with first one star being occulted and then the other.) One of the most spectacular sights here is the Moon passing in front of a bright star cluster such as the Pleiades or Hyades; at such times you can not only watch it ponderously moving through space (especially when its dark part is faintly illuminated by *earthshine*), but also see it seemingly suspended three-dimensionally against a backdrop of more distant stars!

Precise timing of occultations provides valuable information on such topics as the profile of the lunar limb, the Moon's orbital position and motions and, in the case of asteroids, their sizes and shapes. Especially dramatic are *grazing occultations*, where objects are seen going in and out of valleys and behind mountains at the edge of the Moon. Adding to the fun is a slow "rocking" or "nodding" of the Moon in both longitude and latitude known as *libration* (which allows us to see some of its back side, or a total of 59% of the entire Moon). As a result, features on its edge are constantly changing their aspect to us. Among others, the US-based International Occultation Timing Association (IOTA) collects and analyses observations of lunar occultations. It can be reached at www.occultations.org. Among other important organizations that collect all types of observations of the Moon (as well as those of the planets and other Solar System objects) by amateur astronomers are the historic British Astronomical Association (BAA) and the US-based Association of Lunar and Planetary Observers (ALPO). They can be contacted at www.britastro.org (or by e-mail at office@britastro.com) and www.lpl.arizona.edu/alpo, respectively.

Two other areas of lunar observing deserve mention, each having aesthetic rather than scientific value. One is that of *lunar eclipses*, in which the Moon passes through the inner and/or outer parts of the Earth's long cone-shaped shadow – turning pink, copper, orange, rose, or even dull- or blood-red due to light being refracted by the Earth's atmosphere into the shadow. Here again, the slow continuous motion of the Moon eastward through the shadow is strikingly obvious even in binoculars. Note also that the shadow of the Earth projected onto the Moon is curved during all phases of the eclipse – vivid proof (known even to the ancients) that our planet is round! The other area is that of *conjunctions* – or the close coming together in the sky – of the Moon with the brighter stars and, especially, brilliant planets such as Venus or Jupiter. A typical sight is the lovely crescent Moon hovering over the western horizon at dusk (or over the eastern horizon at dawn), its dark portion beautifully illuminated by earthshine and accompanied by one or more brilliant planets and/or stars. Such gatherings are very striking, especially viewed in binoculars and rich-field telescopes (RFTs). On occasion, both the Moon itself and a planet or bright star can be encompassed within the same low-power eyepiece field of standard telescopes, offering a stunning contrast between different kinds of celestial objects.

There's no greater pleasure than embarking on a nightly telescopic "sightseeing tour" of the Moon's surface features through a small telescope, using a good lunar map to guide you. Wide-angle, low-power views make our satellite seem suspended in space, while high-power ones give the strong impression of hanging in lunar orbit! Being an extended object and so bright, the Moon takes high magnification well, given good seeing conditions. Dividing the distance of the Moon by the power used will tell you how far away it appears in the eyepiece. Thus, a magnification of 240× will seemingly bring it to within 1,000 miles at its average distance from us. The lunar detail visible in a good 6-inch aperture telescope (of any type) at this power is simply staggering! Sky Publishing offers several excellent lunar maps that are ideal for such a sightseeing tour. Note here that if you are using a star diagonal (which is nearly always the case) on a refractor, Cassegrain-style reflector, or catadioptric telescope, a traditional inverted lunar map will be all but impossible to compare with the real Moon, because of the mirror image produced by the diagonal. Sky Publishing also offers lunar maps that compensate for this. On-line and printed copies of their latest catalog (which contains many other valuable observing guides and references) can be obtained at www.skyandtelescope.com. Two other excellent lunar works are *Atlas of the Moon* by Antonin Rukl and *The Modern Moon: A Personal View* by Charles Wood. The latter author is also originator of the popular *Lunar 100 Card*, which lists the top lunar sights in order of increasing difficulty on one side and has an identifying map of the Moon on the other. All three of these observing aides are by Sky Publishing (2004).

Planets

We begin with the elusive planet *Mercury*, closest of them all to the Sun. Like the Moon, this tiny orb goes through *phases*, never appearing larger than about 7 arc-seconds in apparent size when best placed for viewing. This occurs at its greatest *elongations* east (in the evening sky) or west (in the morning sky) of the Sun, at which time it appears half-illuminated. A 3-inch scope at 100× will show the phases and its tiny pinkish disk, while 200× or more in 6-inch and larger apertures reveals hints of subtle surface markings. Unfortunately Mercury never strays far from the Sun and so is always fairly low above the horizon, where its image is often degraded by the atmospheric turbulence and haze typically found there. For this reason, it is best observed in broad daylight when it's much higher in the sky. But finding it then can present something of a problem without the use of precision setting circles – especially with the dazzling Sun so dangerously close by. Go-To or GPS-equipped scopes that have been previously aligned on stars at night and not moved should be able to find it readily and safely using their databases.

Next in order of distance from the Sun is *Venus*, the radiant "Evening/Morning Star" and brightest of all the planets – and indeed of all celestial objects aside from the Sun and Moon. Like Mercury, it goes through *phases*, from full when on the opposite side of the Sun from us to a big crescent when passing between it and us. At such times, its minimum distance from us is some 25,000,000 miles – significantly closer than Mars or any other planet ever approaches. Its crescent then spans over 60 arc-seconds in apparent size and can be seen in

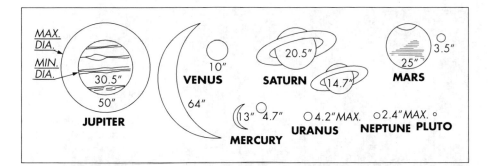

Figure 11.6. The maximum and minimum relative sizes of the planets in arc-seconds to the same scale. Note that Venus is at its apparent largest in the crescent phase, which is why it's then at its brightest rather than when full, as you might expect! Also, remote Pluto remains totally star-like in backyard (and most observatory-class) telescopes, no matter how large they may be.

10 × 50 binoculars. Atmospheric extensions of the *cusps* of the crescent into the nighttime side can also sometimes be seen around this time. At its greatest *elongations* east and west of the Sun when half-illuminated, it averages 24 arc-seconds across. Although Venus is perpetually shrouded in dense clouds, subtle *shadings* (including an elusive atmospheric radial "*spoke system*" that's sometimes glimpsed) and *terminator irregularities* can be seen on occasion in telescopes as small as a 4-inch glass. Especially fascinating are the *ashen light* and the *phase anomaly*. The former is an occasional apparent illumination of the dark side seen when Venus is in the crescent phase, which has been attributed to possible intense auroral activity. The latter is the strange fact that when Venus reaches its predicted half-full phase (or *dichotomy*) as seen from Earth, the terminator is not straight, as it should be. Instead, this occurs several weeks earlier than predicted for evening elongations and several weeks later for morning ones. Sometimes referred to as the Schröter effect after its discoverer, its exact cause is unknown. As with Mercury, observations are best made in daylight, when not only is the planet higher above the horizon, but also the sky background reduces the intense glare of the planet as normally seen at night. Another useful technique here is to observe Venus in the evening or morning twilight, when contrast with the illuminated sky also helps reduce glare and improve image quality.

Continuing to move outward from the Sun we find *Mars*, the famed "Red Planet" (it's actually orange!) of mythology, fiction, and science, once believed to be an abode of intelligent life. At its *oppositions* every 26 months, it can come as close to us as 34,000,000 miles and appear as large as 25 seconds of arc in apparent size. At such times, a magnification of about 75× makes it look as big in telescopes as does the Moon to the unaided eye! Other than at opposition the planet can be a disappointing sight, as it shrinks to well under 10 arc-seconds in size. But during its close approaches, many fascinating features can be seen – especially in 8-inch and larger telescopes.

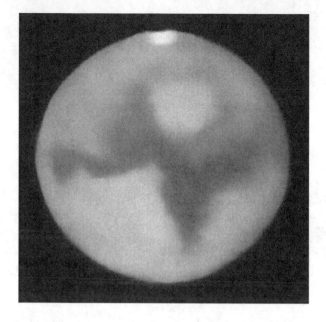

Figure 11.7. A stylized drawing of Mars as typically seen in large backyard telescopes at high power during its close approaches to us every 26 months when at opposition. Shown here are its dark surface features, its lighter deserts, and one of its white polar caps. Note also the streaky linear markings – the famed "canals" reported by Percival Lowell and other observers, both past and present. South is up, as viewed in an inverting telescope.

These include the white *polar ice caps*, the *dark melt band* around each one as they alternately melt with the onset of summer and the accompanying "*wave of darkening*" equatorward, the expansive *orange deserts*, large *dark markings* (which turn from gray in the winter to bluish-green in summer but are not vegetation growing, as was once believed), vast yellowish *dust storms* which occasionally cover much of the planet, bluish-white *clouds*, streaky "*canals*" and the rotation of the planet every 24.5 hours. Of special interest are the occasional *flares* or flashes that have seen around the time of opposition, now believed to be sunlight reflecting from ice deposits on the surface. However, this doesn't explain those seen projecting off the limb of the planet itself. There's also the mysterious *blue clearing* in which the normally opaque Martian atmosphere as seen through blue filters becomes transparent, revealing surface features. Mars has two well-known but very tiny *satellites*, Phobos and Deimos. Shining dimly at about 11th magnitude and rapidly orbiting close to the planet, they require a 10-inch telescope to be seen, in addition to excellent seeing *and* transparency, owing to the glare from Mars itself. (Some observers place the planet behind an occulting bar to hide it while looking for the moons.)

The "Giant Planet" *Jupiter* is the most dynamic and exciting object in the Solar System for most observers. At its favorable oppositions, it comes to within 365,000,000 miles of us and its disk grows to as large as 50 arc-seconds in size, making the planet obviously non-stellar even in 7× or 10× binoculars. Among the features to be seen here with just a 3-inch telescope at 75× to 100× are its *polar flattening* or elliptical shape (due to its rapid *rotation* of less than 10 hours, which is obvious at the eyepiece as features transit the planet's central meridian), colorful dark belts or *bands* and bright *zones*, the *limb and polar darkening*, snake-like *festoons* and cell-like *ovals*, the famed *Great Red Spot* (which currently

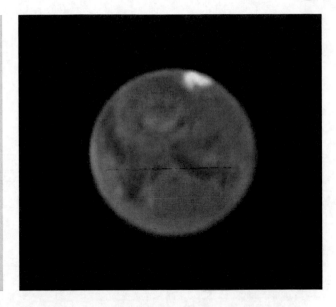

Figure 11.8. Mars as imaged through an 11-inch Schmidt–Cassegrain catadioptric. Compare the surface features seen in this picture with those shown on the drawing in the previous figure (the planet having rotated significantly to the right here). Courtesy of Dennis di Cicco.

Figure 11.9. A dynamic but moonless Jupiter photographed through an 8-inch Newtonian reflector, showing its two prominent equatorial belts. This was one of those rare occasions when *none* of its satellites was visible – all four moons being either in eclipse, transit, or occultation. The rapid rotation of the planet (less than 10 hours) is evident within just minutes of careful watching, and results in Jupiter being noticeably flattened through its poles. As for planets in general, much more detail can be seen visually than captured on film. (However, modern high-speed electronic imaging can now largely capture what the eye sees.) Courtesy of Steve Peters.

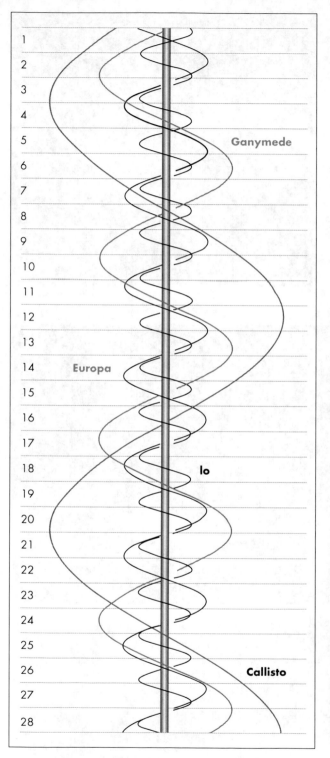

Figure 11.10. The majestic dance of Jupiter's four bright Galilean satellites about the planet over a four-week period. (The double line down the center of the diagram represents the disk of Jupiter itself.) The ever-changing positions of the moons makes this the most fascinating of all the planets for many observers. Visible in steadily held binoculars, they're a thrilling sight in even the smallest of telescopes.

looks salmon-pink), and the thrilling phenomena of its four bright *Galilean satellites*, Io, Europa, Ganymede, and Callisto.

These jewel-like moons are visible in steadily held binoculars (there are even a number of naked-eye sightings in the literature!) and can be seen changing position from night to night as they dance about the planet. Even a 4-inch telescope at 100× on nights of steady seeing will show that the moons are "nonstellar" in appearance, having tiny disks (on which markings have been glimpsed with large amateur scopes). And here we truly have a "three-ring circus" – or actually a four-ring, since there are four satellites! Perhaps most spectacular are the *eclipse* disappearances and reappearances, as the orbiting moons move into or leave Jupiter's huge invisible shadow. Io, being the innermost and fastest-moving of them, requires only a minute or so to fade and then to brighten again an hour or two later, while the slower-moving outermost two moons take a good five minutes to do so. It's an indescribable thrill to see one of these satellites disappear or reappear right at the time predicted! There are also the *occultation* disappearances and reappearances, as the moons pass behind and then exit the planet's disk. Another set of phenomena are the *transits* both of the moons themselves across the face of Jupiter and of their ink-black shadows cast on the cloudtops. And as if that's not enough action, every six years as the Earth passes through the plane of Jupiter's orbit, the moons themselves can eclipse and occult each other in a thrilling series of *mutual satellite phenomena*! Times for all of the foregoing events are given monthly in *Sky & Telescope* magazine and also on its web site. (Incidentally, even though Jupiter is known to have more than 60 moons at the time of writing, none of the others can be seen visually except in the largest of observatory telescopes.)

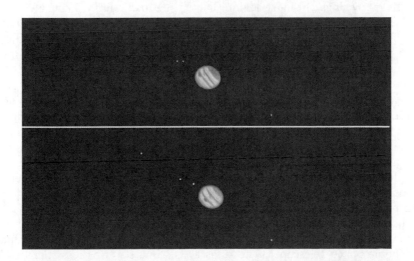

Figure 11.11. Two sketches at the eyepiece of a 6-inch reflector showing the change in relative positions of Jupiter's four bright satellites about the planet over a period of three hours. When one of the inner moons (which move the most rapidly) is near the edge of Jupiter itself or going into its shadow, its orbital motion can be detected in just a matter of minutes!

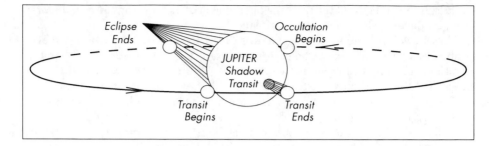

Figure 11.12. This diagram shows the aspects of the eclipses, occultations, and transits of Jupiter's four bright moons, as well as the transits of their shadows on the planet's cloud-tops. These ongoing events make this giant world ever-fascinating to watch.

For most people (whether stargazers or not), the "Ringed Planet" *Saturn* is without question the most beautiful and ethereal of all celestial wonders. At its oppositions, this stunning object passes to within about 750,000,000 miles of us and subtends an apparent size across the rings of some 50 arc-seconds (about the same apparent size as Jupiter at opposition), while the ball of the planet appears about 20 arc-seconds in size. Just 25× on a 2-inch glass shows the disk and rings, looking like some exquisite piece of cosmic jewelry, while the view in 12- to 14-inch apertures at 200× or more leaves one gasping for breath in astonishment! (The planet looks egg-shaped in 10 × 50 and larger binoculars.) These razor-sharp *rings* are made up of billions of chunks of ice and rock orbiting Saturn. There are actually many rings, the most obvious being the broad middle B ring and – separated from it by the black *Cassini Division* – the narrower outer A ring. There's also the dusky inner C ring, most obvious as a shading where it crosses the ball of the planet. The rings cast their *shadows* on the planet's cloud-tops and the planet casts its shadow on the rings behind it. Other features to note are the *polar and limb darkening* and the obvious *polar flattening* of the planet due to its rapid *rotation* of just over 10 hours at the equator. Transient *white spots* are occasionally seen in the atmosphere. On rare occasions occultations of stars by the rings and planet have been observed, one of the most spectacular of recent times being that of 5th-magnitude 28 Sagittarii in July of 1989. At the time of writing, Saturn is known to have some 50 *satellites*, at least four or five of which can be seen in a 6-inch glass. Titan, the largest, is visible in large binoculars and several others can be glimpsed in a 4-inch. From one year to another, Saturn's rings noticeably change their inclination or tilt towards us. Every 15 years, the Earth passes across the ring plane, at which time the rings disappear from our view for several weeks in amateur scopes – and for several hours in observatory instruments, including the Hubble Space Telescope. (The rings are estimated to be less than 100 *feet* thick!) For a year or so before and after these edgewise events (the next of which will occur in September of 2009), the moons can be seen "threading the rings" like beads on a string! It's also near these times of ring crossings that Saturn's moons go through phenomena similar to those of the satellites of Jupiter, as discussed above.

Figure 11.13. Magnificent Saturn with its awesome ice-rings imaged through an 11-inch Schmidt–Cassegrain catadioptric. Even at low power in a small glass, the planet looks like some exquisite piece of cosmic jewelry. As telescope aperture and magnification increase, it becomes a source of ever-increasing wonder! Here the rings are seen moderately wide open. Note the thin, dark Cassini Division encircling the rings, and also the shadow of the planet on the rings behind and just to the left of the ball. Courtesy of Steve Peters.

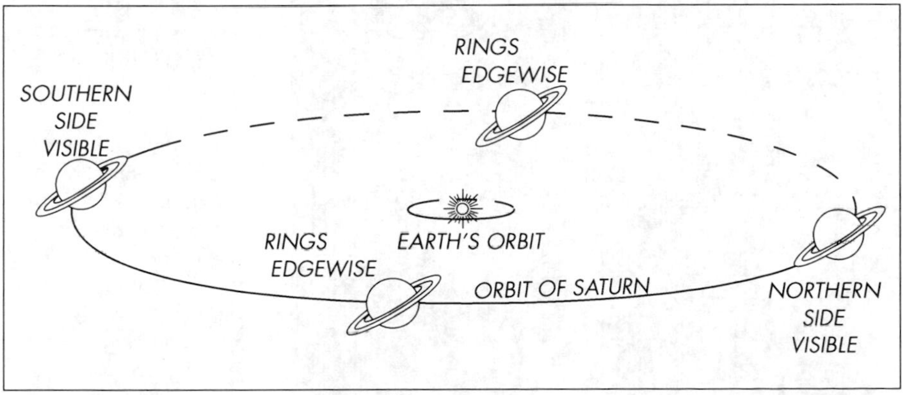

Figure 11.14. The ever-changing aspect of Saturn's rings as seen from Earth resulting from its 27-degree axial tilt. Currently they are slowly closing up, with the next edge-on ring-plane crossing occurring in September of 2009. At that time they will disappear from sight in all but the largest telescopes and the ringed planet will briefly appear ringless.

Uranus was the first of the major planets to be discovered by telescope (by Sir William Herschel in 1781 using a 6.2-inch home-made metal-mirrored reflector). Reaching magnitude 5.5 when at opposition, it's just visible to the unaided eye; you can see it easily in binoculars and follow its slow orbital motion among the stars. Its tiny greenish disk (only 4 arc-seconds in size) is obvious in a 4- or 5-inch glass at 100× on a steady night but is virtually featureless in amateur telescopes. Its retinue of more than 20 dim moons (including its two largest ones, Titania and Oberon) require large telescopes to glimpse visually, although several of the brighter ones have been imaged in apertures as small as 8 inches. Uranus' extremely faint and narrow ring system has only been imaged by spacecraft.

Neptune at 8th magnitude is visible in binoculars as a star-like object, while 5-inch and larger telescopes at 150× will show its tiny bluish disk just 2.4 arc-seconds in apparent size. Its largest moon, Triton, has been glimpsed visually in large amateur instruments, but the remaining dozen or so currently known, as well as its ultra-faint ring system, all lie beyond backyard telescopes.

Figure 11.15.
Comet Hale–Bopp photographed with a 50-mm camera, showing both its blue gas tail and more prominent pinkish-white dust tail in addition to its bright coma or head. This is how it appeared to the unaided eye and through binoculars at its peak during the spring of 1997. Comet tails always point away from the Sun (because of radiation pressure) – which means that when inward-bound they trail the head, but after rounding the Sun and heading back out into the Solar System they go tail-first! Courtesy of Steve Peters.

The outermost planet, star-like *Pluto*, never gets brighter than magnitude 14.5 and to glimpse it at all requires at least a 10-inch telescope, a dark transparent night, and a detailed chart of its position among the starry background. Its largest moon Charon lies less than an arc-second from the planet and has never been seen visually in any telescope. The real thrill in observing the outer planets Uranus, Neptune, and Pluto with your telescope is simply being able to look upon these remote and mysterious worlds lurking in the outer Solar System from across *billions* of miles of interplanetary space.

Comets

Another traditional field of observational astronomy in which amateurs have done outstanding work in the past is searching for and monitoring *comets* – using everything from the unaided eye and binoculars to large backyard telescopes. The real lure here, of course, is the chance to achieve astronomical fame by discovering a new comet and having your name attached to it!

Some noted observers such as Leslie Peltier, David Levy, and William Bradford have discovered *dozens* of them, while others have found only one – but perhaps one so spectacular that it and its discoverer became world-famous. These typically have been the result of purposeful and methodical sweeping of the sky with binoculars and wide-field telescopes. But there have also been totally accidental discoveries, as well as several by the same observer within a night or two of each other! Today, however, sophisticated professional, wide-angle survey telescopes equipped with CCD "eyes" (designed for patrolling the sky for Near-Earth Asteroids – see below) are making most of the finds as a byproduct of their main program. Even so, it's believed that many comets are still missed as they enter the inner Solar System, so opportunities do present themselves for their visual discovery by amateurs. Should you believe you may have found a new comet, you must do the following.

First, using a detailed star atlas (see Chapter 13) carefully check the starfield it lies in for deep-sky objects posing as comets. Second, using the *Sky & Telescope* web site, make sure that it's not an already known new or returning comet. Third, wait at least 15 to 30 minutes to make sure that the object is actually moving – something comets must do! If your suspected comet passes these tests, note its position on the atlas, the direction it's moving, and its apparent brightness (using surrounding stars of known magnitude defocused to match its appearance). Then immediately report your discovery to the International Astronomical Union's Central Bureau for Astronomical Telegrams, located in the Harvard–Smithsonian Center for Astrophysics at 60 Garden Street, Cambridge, Massachusetts 02138. Most discoveries of comets and other transient astronomical objects/events today are usually communicated by e-mail to cbat@cfa.harvard.edu rather than by telegram. (The IAU itself is headquartered in Paris; its official web site can be found at www.iau.org.) A good primer for would-be comet hunters, as well as a comprehensive text covering all aspects of comets (including photographic and electronic imaging, photometry, spectroscopy, and even orbit calculation using computers) is *Observing Comets* by Gerald North and Nick James (Springer, 2003).

Asteroids

It's estimated that there may be more than a million minor planets or *asteroids* larger than a mile across orbiting the Sun between Mars and Jupiter. Ceres, Pallas, Juno, and Vesta were the first four to be discovered (which was done visually with telescopes). Of these, only Vesta becomes bright enough (magnitude 5.5) to be visible with the unaided eye. But hundreds can be seen in binoculars and literally thousands in backyard telescopes. Following their slow movement across the background starfield from night to night constitutes their chief attraction to casual observers. However, most are not round and they change in brightness as they spin, so their rotational periods can be determined by making continuous magnitude estimates (or measurements if done electronically). And as mentioned above when discussing occultations by the Moon, an indication of their shapes can also be obtained by recording the precise times of their disappearances and reappearances from the edge of our satellite. There are even some amateurs who have joined the search by professional planetary astronomers for threatening Near-Earth Asteroids (or NEAs), more than a thousand of which are known to cross the Earth's orbit. This includes several that have actually passed *between* the Earth and the Moon in near misses! The clearing house for asteroid discoveries and observations is the IAU's Minor Planet Center, also located at the Harvard–Smithsonian Center for Astrophysics.

Meteors

"Shooting stars" may not seem to be a class of object that would come under the topic of using telescopes and binoculars. But, in fact, *meteors* are often seen unexpectedly flashing through the eyepiece fields of such glasses – typically sending a rush of surprise and excitement (especially in the case of bright ones!) through the unsuspecting observer. Traditionally, naked-eye studies of meteors have focused on counting hourly rates, and this applies equally to those seen through optical instruments. Keeping track of their numbers over a given interval of time is normally a very casual program that can be done while conducting other observations. But during one of the major annual showers such as the Perseids or the Geminids, the eyepiece field may be flooded with faint streaks of light – especially in large-aperture, wide-field telescopes like Dobsonian reflectors. Most of these typically lie near the limit of vision, but on occasion a bright naked-eye meteor (and its train) may actually pass across your view. Talk about being surprised and thrilled! There are two main organizations collecting meteor observations by amateur astronomers. One is the American Meteor Society (AMS), whose web site address is www.amsmeteors.org. The other is the International Meteor Organization (IMO), which can be reached at www.imo.net.

Before moving on, it should be mentioned here that one of the standard classic references to observing all of the various objects discussed above is *Observational Astronomy for Amateurs* by J.B. Sidgwick. Together with his companion volume, the *Amateur Astronomer's Handbook*, they provide an authoritative and comprehensive guide to visual Solar System (as well as stellar) astronomy for serious amateurs that, while somewhat dated, remains very valuable. Both were reprinted by Dover Publications in 1980 from the 1971 Faber & Faber third edition. And while they have since been updated and reissued several times under revised titles, the original editions of Sir Patrick Moore's two classics, *Guide to the Moon* and *Guide to the Planets* from the early 1950s, provide charming and nostalgic overviews of their subjects. Dating from a time when the Moon and planets were still magical and mysterious places for amateur astronomers to explore in their telescopic spaceships, they are simply "must" reading for cloudy nights – if only you can manage to find copies!

Artificial Satellites

Just as meteors seem unlikely telescopic targets, so too do *artificial satellites*. The wide fields of view and rapid pointing capability of binoculars have long been used to follow satellites across the sky. Their frequent variation in brightness due to spinning (especially in the case of the solar panels on the Iridium satellites, which can catch the sunlight and cause them to flash brighter than Venus for a few seconds!) and their slow fading as they pass into the Earth's shadow-cone are things to watch for using both binoculars and the unaided eye itself. But there are so many of these objects that there's rarely a night when at least several satellites don't also pass through the eyepiece field of a telescope during an observing session (sometimes several in rapid succession!), no matter where it's pointed in the sky. Undoubtedly the most spectacular of them all is the International Space Station (or ISS) – especially when the Space Shuttle is docked alongside it. If they happen to pass through the eyepiece, their characteristic shapes are quite obvious even though the view lasts only a second or so. The two craft have even been seen silhouetted against the Moon (and the Sun)! And they have also been imaged by amateurs, both together and apart, using telescopes on special tracking mountings designed to follow their rapid orbital motions. I am continually amazed at how often a satellite will pass *directly over* prominent deep-sky objects such as the Ring Nebula, the Hercules Cluster, or the Andromeda Galaxy. While such passes are definitely an unwelcome annoyance to astroimagers, they are thrilling sights to visual observers. The National Aeronautics and Space Administration (NASA) offers a number of Internet sites that provide satellite predictions for those interested in viewing these hurtling man-made moons. Probably the best of these is the one known as "J-Pass," which can be accessed by placing the following address in your web browser: http://science.nasa.gov/Realtime/JPass/ PassGenerator. You simply enter your e-mail address in the appropriate box on the site's home page to subscribe. J-Pass will provide times and directions in which to look for passages over your location of up to 10 satellites, including the ISS and the Space Shuttle (when flying).

CHAPTER TWELVE

Stellar System Observing

First-Magnitude and Highly Tinted Single Stars

Technically, any object lying beyond the confines of our Solar System is a *deep-sky object*. However, here we reserve that category for such remote denizens of space as star clusters, nebulae, and galaxies, which are covered as a separate class in Chapter 13. Data to help in observing many of the objects mentioned in both chapters will be found in the Showpiece Roster in Appendix 3, and also in the various star atlases and references mentioned in the text itself.

Before proceeding, I should like to share with you a quote from Charles Edward Barns' long-out-of-print classic, *1001 Celestial Wonders*. Its poetic mandate is appropriate for all classes of objects viewed by stargazers, but especially so for the wonders discussed in this and the following chapter:

> Let me learn all that is known of them,
> Love them for the joy of loving,
> For, as a traveler in far countries
> Brings back only what he takes,
> So shall the scope of my foreknowledge
> Measure the depth of their profit and charm to me.

We begin with single suns, and in particular the very brightest stars – those having an apparent visual magnitude of +1.5 or brighter. They are members of the so-called "First-Magnitude Club," and there are 23 such luminaries in this exclusive group scattered over the entire sky. They range from dazzling Sirius (α Canis Majoris) at −1.4 (the brightest of all stars after the Sun itself) to Adhara (ε Canis Majoris) at +1.5. Of these, 16 are visible from mid-northern latitudes – or 17 if we count Canopus (α Carinae), which is the second-brightest star at magnitude −0.6 and is visible from Florida and Texas.

Aside from a very few of the first-magnitude stars that happen also to be attractive visual double or multiple systems, their principal charm lies in their lovely heavenly hues. As a class, these stellar jewels are bright enough for their colors to be perceived directly even with the unaided eye, while binoculars or small telescopes focus more than enough photons onto the retina to make their hues unmistakable even to beginning observers. If you think all stars are simply white, just compare the color of ruddy Betelgeuse (α Orionis) with that of bluish Rigel (β Orionis) in the well-known winter constellation of Orion – or the blue-white hue of Vega (α Lyrae) with that of golden Arcturus (α Bootis) in the summer sky! As J.D. Steele pointed out over a century ago, "Every tint that blooms in the flowers of Summer, flames out in the stars at night."

For me, one of the great joys of leisurely stargazing is to go outdoors after sunset on a clear night, with binoculars in hand, and watch these colorful luminaries slowly make their appearance as the sky darkens. It was surely this simple activity that Henry Wadsworth Longfellow (himself a stargazer) had in mind when he penned:

> Silently one by one, in the
> infinite meadows of heaven,
> Blossomed the lovely stars,
> the forget-me-nots of the angels.

Another aspect of observing the brightest stars that's directly related to their perceived colors concerns their spectral types (both characteristics being determined by the temperature of the star's atmosphere). Using an eyepiece star spectroscope (see Chapter 7) on a telescope of 4-inch or more in aperture, it's possible to view the dark absorption lines and bands (and in some cases, emission lines if present) clearly enough to distinguish the various spectral classes. The familiar spectral sequence runs from types O, B, and A at the hot end, through warm F and G types, to K, M, R, N, and S cool ones – all relatively speaking, of course!). The cooler a star is, generally the more absorption lines that will be visible, and in the very coolest stars entire bands will be seen. (Readers interested in learning more about spectroscopes and stellar spectroscopy should consult Mike Inglis' excellent *Observer's Guide to Stellar Evolution*, Springer-Verlag, 2003.)

It turns out that some of the most highly tinted stars in the sky are not among the brightest ones but rather lie near or below naked-eye visibility because of their distances. Among the more famous of these are Herschel's Garnet Star (μ Cephei), La Superba (Y Canum Venaticorum), and Hind's Crimson Star (R Leporis). These are all pulsating red supergiant suns, slowly changing in brightness over time, like colossal cosmic beating hearts! (See the section on variable stars later in this chapter.) While some of these ruby gems can be seen in binoculars, a 3- to 6-inch telescope shows many more of them scattered about the sky. One of the reddest stars to be found is T Lyrae, located near Vega in Lyra and set against the rich backdrop of the summer Milky Way. In 8-inch and larger glasses, it's truly a stunning sight, looking for all the world like some interstellar traffic light far out in the depths of space! In her classic work of a century ago, *The Friendly Stars*, Martha Evans Martin tells us that "The stars we love best are the ones into whose faces we can look for an hour at a time, if our fancy so leads us." As a class, the stars we've discussed in this section provide perfect opportunities for doing just that. Some of the most attractive of these stellar gems are listed in the Showpiece Roster in Appendix 3.

Double and Multiple Stars

We begin here by defining exactly what double and multiple stars are: two or more suns placed in close proximity to each other in the sky as seen with the unaided eye, binoculars, and/or especially telescopes. With the exception of stars that just happen to lie along the same line of sight but are actually far apart in space (known as *optical doubles*), these objects are physically (gravitationally) bound together as a system. In some cases, they are separated sufficiently to be simply drifting through space together as *common-proper-motion* pairs, while in others they are orbiting around the common center of gravity of the system as a true *binary*.

There are many types of binaries, the actual classification depending on how close together the components are and what type of instrumentation is required to see them. Of primary interest to amateur astronomers are the *visual binaries*, those resolvable in backyard telescopes. These have orbital periods ranging from a few decades to many centuries, and angular separations of anywhere from around half an arc-second to several minutes of arc. (Of the various other types, *spectroscopic* and *interferometric binaries* typically have much shorter orbital periods – in some cases, just *hours*!) Throughout this section, the term "double star" is taken to mean both visual double *and* multiple systems.

Double stars are the tinted jewels and waltzing couples of the night sky. Their truly amazing profusion and seemingly infinite variety of colors, brightness, separations, and component configurations make them ever-fascinating as both objects of study and targets for leisurely exploration with the telescope. Astronomers estimate that at least 80 percent of the stellar population exists as pairs and multiple groupings. Abounding among the naked-eye stars, they are available to even the smallest of instruments for viewing on all but the worst of nights – even in bright moonlight, and through haze and heavy light pollution. Literally *thousands* of them lie within reach of even a 2- or 3-inch glass!

After the Moon and planets, double stars are typically the next target for the beginning stargazer before jumping into the fainter and more distant realm of star clusters, nebulae, and galaxies. And indeed they should be, for not only are they bright and easily found, but these are truly exciting objects in themselves. The vivid hues of bright, contrasting pairs are sights never to be forgotten. There's the magnificent topaz and sapphire Albireo (β Cygni), the vivid orange and aquamarine-blue of Almach (γ Andromedae), and the red and green of Rasalgethi (α Herculis). Brilliant blue-white pairs such as Castor (α Geminorum), Mizar (ζ Ursae Majoris), and Rigel (β Orionis) look like glittering celestial diamonds

Figure 12.1. A magnificent color image, taken with an 11-inch Schmidt–Cassegrain catadioptric, of Albireo (β Cygni) – considered the most beautiful double star in the sky and seemingly everyone's favorite stellar combo! Its exquisite topaz-orange and sapphire-blue hues are unmistakable even in a 2-inch glass at 25× and they stand up well in the largest of telescopes (which often saturate colors from *too much light* and also unduly separate the stars). This lovely pair can be resolved in steadily held 10 × 50 binoculars. Here truly is one of the grandest sights in the entire heavens. Courtesy of Johannes Schedler.

against the black velvet of space. And there are stunning multiple systems such as the Double-Double (ε Lyrae), the Trapezium (θ¹ Orionis), and Herschel's Wonder Star (β Monocerotis).

As in other areas of amateur astronomy, part of the joy of observing double stars is in sharing views of these lovely objects with others. Their unsuspected beauty typically brings gasps of astonishment and delight from

Figure 12.2. An eyepiece impression in black & white through a 4-inch refractor of the colorful double star ι Cancri, which is located above the naked-eye Beehive Cluster in Cancer. The size of the star images have been exaggerated for clarity (simulating the effect of less than ideal seeing conditions!) – on steady nights the components actually appear as radiant stellar pinpoints. Nicknamed the "Albireo of Spring" by the author, its lovely orange and blue tints remind many observers of that famous pair. South is at the top, as seen in an inverting telescope.

first-time observers as they peer into the eyepiece. Many double-star aficionados can trace their lifelong love affair with these stellar combos to just such an encounter. Seeing sights such as those listed above often draws the observer into deeper study of double stars. Some people, desiring to see as many of them as possible, will embark on a "sightseeing tour" of the night sky using lists like those given in Appendix 3, in the references below, and at the end of this chapter. Others, wanting to capture these sights permanently, may do so by drawing what they see at the eyepiece – or by attempting to image it photographically or electronically using CCD or video cameras.

Given time and patience, the orbital motions of bright pairs such as Porrima (γ Virginis), ξ Ursae Majoris, and Castor itself become evident over a period of years. The sight of two distant suns slowly dancing about each other in the depths of interstellar space is a thrill quite beyond words! This may lead the observer to begin regular measurements of binary stars with a micrometer or other such device to follow their movements. This is not only a fascinating activity in itself, but one that also happens to be of great value to professional astronomers

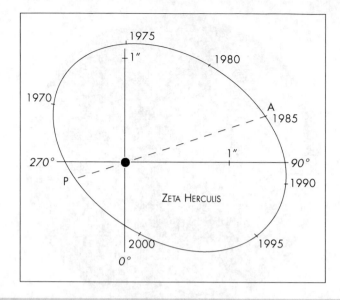

Figure 12.3. The apparent orbit of a typical fast-moving visual binary – in this case ζ Herculis, one of the corner stars in the Keystone asterism of Hercules. Having a period of just 34 years, the author has dubbed this object "Herschel's Rapid Binary" after its discoverer. The companion has now made over *six complete circuits* of the primary since Sir William found it in 1782! "A" represents *apastron*, or maximum separation of the stars in the true orbit, while "P" stands for *periastron* or their closest approach to each other. This tight, dynamic pair can be resolved in a good 4-inch glass when at its widest, which last occurred in 1991. It was at minimum separation in 2001 and is now opening up again.

studying the orbital dynamics of double-star systems. Binaries provide our only means of directly measuring the masses of the stars – information crucial to the study of the birth, evolution, and death of stellar systems. The world clearing house for reporting double-star measurements is the United States Naval Observatory in Washington, DC. It maintains the monumental *Washington Double Star Catalog* (or *WDS*), containing data on nearly 100,000 pairs and updated continuously. It can be accessed on-line on the Observatory's Internet site at http://ad. usno.navy.mil/wds (along with a veritable galaxy of information on double stars in general). Also, the UK-based Webb Society (named in honor of T.W. Webb, author of the classic work *Celestial Objects for Common Telescopes* – later editions of which contained over 3,000 double and multiple stars) has a very active double-star section that's open to both amateur and professional astronomers. Volume I of its *Webb Society Deep-Sky Observer's Handbook* series provides much useful information on these objects. The Society can be reached at www.webbsociety. freeserve.co.uk.

Of the many star atlases available to double-star observers today (nearly all of which now use a standard symbol for double stars – a line or bar through the

Figure 12.4. An eyepiece drawing from a view at very high magnification through a 13-inch refractor showing the amazing Trapezium multiple-star system (θ¹ Orionis) lying at the heart of the Orion Nebula. A 2-inch glass readily resolves the four brightest members, but at least a 4-inch is required to glimpse the two fainter ones, and an 8-inch is needed to make them obvious. These six suns are the core of a star cluster in formation, condensing out of the nebulosity itself. North is up, as seen when using an erecting star diagonal on the telescope.

star-dot indicating its multiple nature), two stand out. One is the well-known classic, *Norton's Star Atlas* by Arthur Norton, first published in 1910 and now in its 20th edition (Pi Press, 2004) edited by Ian Ridpath. It has excellent maps covering the entire visible sky throughout the year, and accompanying lists of interesting doubles. Unfortunately, beginning with the 18th edition, the valued double-star designations by the various discoverers such as the Struves, the Herschels, and Burnham that once identified the pairs on the maps were dropped. Thus, any of the editions from the 17th or earlier (the more recent of them being by Sky Publishing) are preferred for double-star work – if only the observer is fortunate enough to find a copy on the used market! And there's the magnificent deluxe color edition of *Sky Atlas 2000.0* by Wil Tirion and Roger Sinnott (Sky Publishing, last printing 1998) that plots over 2,700 deep-sky objects and more than 81,000 stars – many of them double and multiple systems – down to visual magnitude 8.5. (It also comes in black-and-white desk and field editions.) The accompanying *Sky Catalogue 2000.0, Volume 2* (Sky Publishing, 1985) provides data on more than 8,000 of the brighter pairs that appear on the atlas itself.

If you are interested in exploring further all aspects of double-star observing, you may want to refer to two of my own publications. One is *Double and Multiple Stars and How to Observe Them* (Springer, 2005), which lists 500 pairs for exploration. The second is *Celestial Harvest: 300-Plus Showpieces of the Heavens for Telescope Viewing & Contemplation* (Dover, 2002). Nearly half of the objects described in it are striking double and multiple stars suitable for observing with telescopes in the 2- to 14-inch aperture range. Many spectacular pairs will also be found in the listing appearing in Appendix 3 of this book. For those seriously interested in measuring double stars (and thereby joining ranks with the few professionals working in this field), *Observing and Measuring Visual Double Stars* (Springer, 2004) – edited by the Webb Society's Bob Argyle – details everything an observer needs to know.

Variable Stars, Novae, and Supernovae

Variable stars are those whose light output varies with time, in periods that range from just minutes to several years. Most of these restless suns are red giants or supergiants, which pulsate not only in brightness but also in physical size. Some, like Betelgeuse, are found among the bright naked-eye stars, while others are at the limits of the largest amateur telescopes. It's conservatively estimated by astronomers that at least 10% of all stars are variable in nature.

Of the several dozen types that have been classified, some of the more interesting variable stars for amateur observers are the following.

Long-period variables are those whose light output changes over a period of months or years with amplitudes of at least several magnitudes. The prototype of this class is the red giant Mira (o Ceti). Called "The Wonderful" or the "Wonder Star" by the ancients who saw its obvious variability, it reaches nearly 2nd magnitude at maximum and fades to well below 9th magnitude (sometimes even as faint as 11th) at minimum over a period of about 330 days. Another well-known member of this class is R Leonis, or "Peltier's Star" as some call it. This object varies between 4th and 10th magnitude over a period of 331 days, with its ruddy hue being obvious in binoculars in its brighter stages and remaining a lovely sight in a 6-inch telescope even at minimum light. (Many of the long-period variables display unmistakable and often striking red hues throughout their cycles.)

The *semi-regular variables* somewhat resemble the Mira stars but have smaller ranges in brightness and irregular periods. Certainly the brightest example is the star Betelgeuse mentioned above (which is the only marked variable of *any* type among the 1st-magnitude stars). It slowly and unpredictably varies in apparent luster from magnitude 0.4 to 1.3 in an approximate period of 5.7 years. Another specimen is the primary of the beautiful double star Rasalgethi (α Herculis), which varies between 3rd and 4th magnitude in no definite or repeatable cycle.

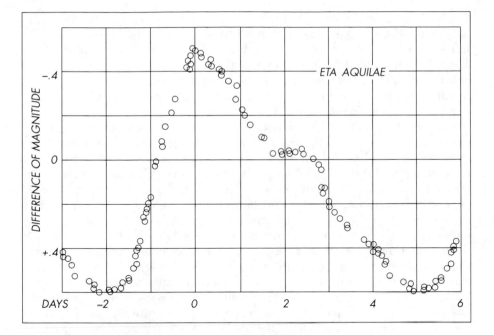

Figure 12.5. A light curve made from visual magnitude estimates of the well-known variable star η Aquilae, a Cepheid having a regular period of just over 7 days. Its changes in brightness from magnitude 3.5 to 4.4 are obvious both to the unaided eye and through binoculars by comparing it with nearby 3.7-magnitude β Aquilae, which shines with a constant light.

There are also the really maverick *irregular variables*, which have small magnitude ranges and completely erratic periods; no such variable is obvious to the unaided eye.

A class of major importance to astronomy and cosmology is that of the *Cepheid variables*, named after the prototype δ Cephei (itself a lovely double star). This sun varies in brightness from magnitude 3.5 to 4.4 in a precise period of 5.4 days, the change being obvious in binoculars and even to the unaided eye. These stars are sometimes referred to as *regular variables* because their light cycles are so constant and repeatable. It's long been known that the greater the period of a Cepheid, the brighter it is in terms of intrinsic or true luminosity (using that of the Sun as a standard). Not only are Cepheids found within the general stellar population and star clusters, but they can also be seen (in large professional telescopes only!) in the nearer spiral galaxies. Comparing their intrinsic magnitudes with their apparent magnitudes provides a valuable yardstick for determining the distances to these objects.

The *eruptive variables* are those that normally shine at a constant brightness and then suddenly without warning brighten up by many magnitudes. One of the best-known examples lies in the constellation Corona Borealis, or the Northern Crown. T Coronae Borealis (also known as the "Blaze Star") usually hovers around 11th magnitude, near the limit for a 3-inch glass. But in 1866, and again in 1946, it suddenly brightened to 2nd magnitude, becoming an obvious naked-eye star at the edge of the Crown asterism. (Interestingly, at the other extreme is the star R Coronae Borealis, located close to T itself. It normally shines just above naked-eye visibility at magnitude 5.7. But in 1962, in 1972, and again in 1977 it disappeared from sight, dropping to nearly 15th magnitude, leading to its popular name as the "Fade-Out Star" or "Reverse Nova.")

Three related groups of highly unstable suns are the *cataclysmic variables*, the *flare stars*, and the *dwarf novae*. These objects hover at very faint magnitudes (typically 14th and below) most of the time, but then suddenly and unpredictably increase in brightness a hundredfold or more. One well-known, spectacular example is the star U Geminorum, which is usually found slumbering between 14th and 15th magnitude but frequently flares to as bright as 8th in little more than 24 hours. Actual *novae* behave similarly but are much more violent, having magnitude increases of thousands of times when they "blow." Some of these have rivaled the brightest stars in apparent brightness at maximum and remained visible to the eye for many months. Over the years, amateur astronomers have played a major role in the discovery of such stellar outbursts. Most sightings of these apparently "new" stars have been made with no equipment other than the unaided eye or binoculars – and an intimate familiarity with the star patterns of the various constellations. (Some of the fainter, more distant novae have also been discovered telescopically by alert visual observers.)

The ultimate in cataclysmic stars are the *supernovae*. Here, if one occurs in our own Galaxy, a previously invisible sun suddenly within hours rises into the negative magnitudes, outshining every object in the sky except for the Sun and Moon (and in at least one case being bright enough to be seen in broad daylight!). Only five have been seen in our Galaxy within recorded history (but they appear frequently in other galaxies, as discussed in the next chapter). The three most famous are Tycho's Star of 1572, Kepler's Star of 1604, and before them the Chinese "Guest Star" of 1054 that produced the Crab Nebula supernova remnant in Taurus. It's now been over four centuries since the last supernova and most astronomers agree that we are long overdue for another one. An amateur astronomer like yourself, scanning the skies nightly, could well be the first to see and report it to the world, thereby achieving – along with Tycho and Kepler – immortal fame. If you think you've spotted a nova (or even better, the next supernova!) on the rise, contact the International Astronomical Union's alert hotline immediately to report it, using the information given in Chapter 11 on discovering comets.

There's also a fascinating class of variable stars that are not really variable at all! These are the *eclipsing variables* – also known as *eclipsing binaries*. As the latter name implies, the variations in brightness of these objects are due to eclipses caused by orbiting companions, rather than to pulsations of size and brightness within the stars themselves. The best-known example in the sky is Algol (β Persei), the "Demon Star" (so named by the ancients, who saw it "winking" at them!). Every 2.9 days – as regular as clockwork – its brightness drops from magnitude 2.1 to 3.4. There are actually two minima. The primary drop that's obvious to the naked eye occurs when a large, dim star in the system passes in front of and partially eclipses a smaller but brighter one; there's also a much less noticeable secondary eclipse half an orbit later as the bright star passes in front of the dim one, reducing its light slightly. The primary eclipse itself last 10 hours. As with other variable stars, Algol's change in brightness can easily be followed by comparing its light to that of a nearby star of constant brightness – in this case 2nd-magnitude Mirfark (α Persei).

Another naked-eye eclipser is β Lyrae, near the famed Ring Nebula in Lyra. Here the light varies continuously from magnitude 3.3 to 4.3 over 13 days because the huge, egg-shaped twin suns continually eclipse one another and are so tightly embraced that their outer atmospheres are nearly in contact. It's truly amazing to think that when we look at stars like Algol or β Lyrae we're actually witnessing the mutual revolution of two suns about each other from across the depths of interstellar space, using no instrument other than the human eye itself! And while these two are bright, naked-eye examples, many other fainter eclipsing systems lie within reach of binoculars and small telescopes.

Variable stars can be viewed for pleasure, and/or observed more seriously with the hope of advancing our knowledge of them by making and submitting magnitude estimates of selected targets. This has traditionally been done by eye with binoculars or telescopes, estimating a star's visual magnitude on a given date by comparing its brightness to that of surrounding stars of known and constant magnitude – typically to an accuracy of a few tenths of a magnitude. Professional astronomers, along with some advanced amateurs, use photoelectric photometers – precision instruments capable of measuring the brightness of stars to within a few hundredths or even thousandths of a magnitude. The US-based International Amateur–Professional Photoelectric Photometry Association (IAPPP) coordinates such cooperative work between the two levels of astronomers and may be reached by contacting its director at douglas.s.hall@iapp.vanderbilt.edu. A recent development along these lines is the increasing use of CCD or video imaging to record the magnitudes of variable stars, as well as those of other celestial objects. (See Chapter 7 for more about these electronic devices.)

Making magnitude estimates of variable stars is of no value to our science unless they are reported to a recognized agency for collection and analysis. One of the world's premier organizations for such work is the American Association of Variable Star Observers (or AAVSO), based in Cambridge, Massachusetts. Founded in 1911 by William Tyler Olcott, author of *Field Book of the Skies* and other observing classics, it's an international clearing house for the submission of variable-star observations by amateur astronomers. The Association provides its members with carefully prepared star charts of objects currently needing observation. Its database contains millions of magnitude estimates, and light curves of countless numbers of variables of all types, which are frequently consulted by professional astronomers working in this field. It also coordinates the monitoring of unpredictable and unstable stars such as the cataclysmic variables and dwarf novae for possible flare-ups. The Association can be contacted at www.aavso.org or by e-mail to aavso@aavso.org. As mentioned above, positive sightings of outbursts should also be reported immediately to the International Astronomical Union's discovery clearing house, since time is critical if transient events like these are to be studied by the professional observatories. An even older organization that also collects magnitude estimates for variable stars is the well-known British Astronomical Association (founded in 1890), already mentioned in Chapter 11 in connection with reporting observations of Solar System objects.

In addition to *Norton's Star Atlas* and *Sky Atlas 2000.0*, mentioned in the earlier discussion about double stars, there's also *The AAVSO Variable Star Atlas* by C.E. Scovil (AAVSO, 1990), explicitly designed for finding and making magnitude estimates of variables. The standard professional compilation of variable stars is the *General Catalogue of Variable Stars* by P.N. Kholopov and B.V. Kukarkin. Published and regularly updated by the Soviet Academy of Sciences, this work is difficult to obtain other than in the research libraries of professional observatories.

Norton's has lists of over 500 variable stars accompanying its star maps, while *Sky Catalogue 2000.0 Volume 2* contains data on more than 2,400 of them (those having maxima brighter than magnitude 9.5) plotted on the *Atlas* itself. Among the many available books on observing these restless suns, two excellent ones are *Observing Variable Stars: A Guide for the Beginner* by David Levy (Cambridge University Press, 1998) and *Observing Variable Stars* by Gerry Good (Springer, 2003).

CHAPTER THIRTEEN

Deep-Sky Observing

Stellar Associations and Asterisms

Jumping now "into the great beyond" of deep space, we begin with two types of typically very large and scattered collections of stars. Though generally unrelated, they both require very wide fields of view to be seen to advantage, giving binoculars and RFTs a definite edge over normal telescopes. *Stellar associations* are physical systems loosely bound by their mutual gravitation, much like star clusters themselves (see below) but covering many degrees of sky rather than many minutes of arc, and much less concentrated. One of the brightest, best-known, and most spectacular of these is the Alpha Persei Association – a radiant splash of stellar gems surrounding the 2nd-magnitude star Mirfak in the constellation Perseus, visible in the skies of fall and winter. The view here in binoculars on a dark night is truly spectacular!

The other type of loose gathering is known as an *asterism*. Asterisms are very distinctive patterns of stars that in most cases are physically unrelated – simply chance alignments similar to optical doubles. But what is probably the most famous asterism of them all turns out to be an actual association or moving cluster of stars as well. We're speaking about none other than the Big Dipper. (Novice stargazers – and a few veteran ones as well! – often refer to the Dipper as a constellation. But in reality, it's only *part* of a constellation – that of Ursa Major, the Great Bear of the sky. Appendix 2 gives a complete listing of the 88 officially recognized constellations.) Another huge and distinctive asterism is the Summer Triangle. It's formed by the three brilliant blue-white luminaries Vega in Lyra, Deneb in Cygnus, and Altair in Aquila. (Cygnus itself contains the well-known Northern Cross asterism.) There are many smaller asterisms to be seen,

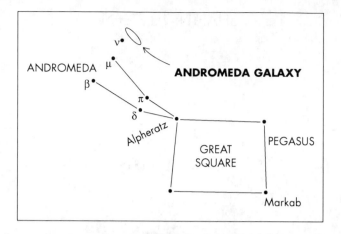

Figure 13.1. Using one of the largest naked-eye asterisms (the Great Square of Pegasus) to find the most magnificent galaxy in the sky (aside from the Milky Way itself!) – that in Andromeda (M31). A diagonal drawn from Markab across the Square to Alpheratz and extended by roughly its own length brings you to the vicinity of this great spiral. Readily visible to the unaided eye on a dark moonless night, it's a binocular wonder and a glorious sight in any and all telescopes.

and perhaps the most amazing of all these "connect-the-dot" patterns is the so-called Coathanger asterism in Vulpecula. Also known as Brocchi's Cluster after its discoverer and officially cataloged as Collinder 399, it consists of six stars in a nearly straight line with four more curving away from the center forming a hook. In binoculars, it looks for all the world like some starry coathanger hanging upside-down in the sky!

Star Clusters

More compressed in size and richer in membership than associations are the *open clusters*, containing anywhere from a few dozen to many hundreds of stars – all gravitationally bound and moving through space together as a commune. Over a thousand are known, most of them lying near the plane of our Milky Way and therefore found mainly in the summer and winter skies, when our Galaxy rides high across the heavens. Some of the brightest of these so-called "galactic clusters" will be found in the famed list compiled by Charles Messier in the late 1700s known as the *Messier Catalogue*, containing 109 entries carrying the prefix "M." Many others are included in the monumental *New General Catalogue of Nebulae and Clusters of Stars* (NGC) published in 1888 by J.L.E. Dreyer – and also in its two subsequent *Index Catalogue* (IC) extensions. Together, these works list over 12,000 objects! A shorter modern roster of clusters, nebulae, and galaxies is that of the *Caldwell Catalog*, compiled in 1995 by the well-known

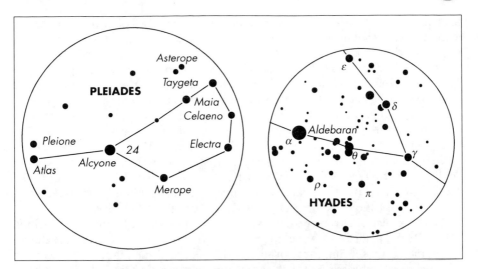

Figure 13.2. The sky's two brightest and most spectacular naked-eye star clusters – the Pleiades (shown on the left as seen through a wide-field telescope) and the Hyades (shown on the right as it appears in binoculars). Amazingly, these glittering stellar jewelboxes lie near each other in the very same constellation of Taurus.

Figure 13.3. The radiant Pleiades star cluster (M45) photographed with a 108-mm apochromatic refractor, its hot blue-white suns shining like diamonds against the black velvet of space. The stars are seen here enmeshed in wispy bluish reflection nebulosity (the brightest part being known as Temple's Nebula, NGC 1435). While discovered visually with a 4-inch telescope, this stellar haze is not easy to glimpse – part of the problem being that the observer's own breath tends to produce a fog surrounding all the stars on cold winter nights! Courtesy of Steve Peters.

British observer and astronomy popularizer, Sir Patrick (Caldwell-)Moore. As with the Messier list, it contains 109 objects – but unlike it, the Caldwell list covers the entire visible sky all the way from the North to the South Celestial Pole.

As with double stars, no two star clusters look exactly alike in appearance. They range from big, glittering stellar jewelboxes spanning several degrees, such as the famed Pleiades and Hyades star clusters in the constellation Taurus (best seen in binoculars and wide-field telescopes because of their large apparent angular sizes), to magnificent sights well under a degree across such as the Wild Duck Cluster (M11) in Scutum and Lassell's Delight (M35) in Gemini (which are at their radiant best in medium-aperture amateur telescopes). And speaking of double stars, there are also double clusters, including *the* Double Cluster (NGC 869 and NGC 884) itself – a magnificent binary swarm of colored gems in Perseus. The visual aspect of such objects never fails to delight the observer.

There are also many surprises – such as the tiny, ring-like planetary nebula NGC 2438 suspended in front of the cluster M46 in Puppis, or the remote cluster NGC 2158 dimly shining through the outskirts of M35 mentioned above, or the faint nebulosity enmeshing the Pleiades, among other clusters. Added treats are the orange or red stars often seen near the center of open clusters, most of whose members are blue-white hot suns. These are cool stars that have evolved off the

Figure 13.4. The rich open star cluster M37 in Auriga, as it appears in high-powered binoculars and small telescopes. This lovely assemblage is the best of the three Messier clusters residing in this constellation (the other two being M36 and M38), all of which are near each other in the sky. This image was taken with a 300-mm f/2.8 camera. Courtesy of Steve Peters.

Figure 13.5. The spectacular Double Cluster (NGC 869 & NGC 884) in Perseus, photographed with a 300-mm f/2.8 camera. A lovely sight even in binoculars, this pair of stellar communes is striking in the smallest of telescopes. Larger apertures show many colored gems amid the predominantly blue-white diamonds present. Amazingly, the two clusters apparently are physically (gravitationally) related, lying at distances of 7,200 and 7,500 light-years, and may be slowly "orbiting" each other! Courtesy of Steve Peters.

Main Sequence into the red-giant stage – vivid proof of stellar evolution right before our eyes! All this and more is available to the devoted observer of open clusters, using just a 3- or 4-inch glass on a dark, transparent night.

Much larger concentrations of stars are found in the *globular clusters*. These glittering stellar beehives contain as many as a *million* individual suns. Of the 150 or so that are known, most reside in the vast galactic halo in the outer regions of our Galaxy, the center of which is located in the direction of the summer constellation Sagittarius. As with open clusters, most of the best globulars can be found in Messier's catalog, including wonders such as the famed Hercules Cluster (M13) in the constellation Hercules. This magnificent ball of stars can be glimpsed with the unaided eye on a dark night and looks like a fuzzy star in 10 × 50 binoculars. A 3-inch glass at 45× shows the ball to have a grainy texture or appearance to it (the precursor to actual resolution), while a 6-inch at 150× resolves stars across the entire image. The view in 12-inch and larger instruments is absolutely breathtaking, with shimmering tinted stars seen all the way into M13's dense core. One observer described the scene as like driving through a snowstorm!

Three other gems from Messier's list not to be missed are M4 in Scorpius, M22 in Sagittarius, and M5 in Serpens – all rivaling M13 itself. The brightest (at 4th magnitude – so bright that it's an obvious naked-eye object having a Greek-letter designation), largest (bigger in apparent size than the Moon), and most spectacular globular cluster of them all is undoubtedly ω Centauri (NGC 5139). This million-sun colossus is visible from the southernmost parts of the United States and is a commanding sight when seen from the Southern Hemisphere. Another naked-eye globular for those in that Hemisphere is 47 Tucanae (NGC 104). Like ω Centauri, it's an amazing sight in even the smallest of telescopes. For observing globulars of any size or brightness, nights having *both* good transparency and steady seeing give the best results in resolving these close-knit star-balls. (It should be pointed out here, however, that nights which are useless for looking at faint nebulae and dim galaxies because of haze, moonlight, or light pollution can still allow both open and globular clusters to come "punching through" owing to their stellar natures.)

When gazing into the heart of a great cluster like one of the above, the mind is forcibly drawn to the possibility of other beings inhabiting the myriad planets that must surely exist there. The appearance of their night sky must be nothing short of astounding. It's estimated that many thousands of stars ranging in brightness from that of Venus to that of the Moon would fill their heavens and always be visible, so that there would really be no night at all! Such is the setting for the classic sci-fi short story *Nightfall* by the prolific science writer Isaac

Figure 13.6. The magnificent globular star cluster M13 in Hercules – better known as the Hercules Cluster – as imaged with an 11-inch Schmidt–Cassegrain catadioptric. We see this colossal stellar beehive by light that left there 24,000 years ago when cavemen were still hunting mastodon! Looking like a small cotton-ball in binoculars, a 3- or 4-inch glass begins to resolve its edges, while glittering stars can be seen all the way into its core in 6- and 8-inch scopes. The view of this and similar bright globular clusters in large backyard telescopes is simply astounding. Courtesy of Dennis di Cicco.

Asimov, in which a civilization lives on a planet where it gets dark only one night in 2,000 years. When it does, the sky's filled with stars. As another author, astronomer Chet Raymo, puts it, observing celestial objects is "50 percent vision and 50 percent imagination" – good advice for viewing wonders such as these. Just picture yourself "out there" among all those stars!

Nebulae

There are five distinct types of nebulous clouds of gas and dust that the observer finds within our Galaxy – four of them luminous or shining, and one non-luminous or dark. First we examine the *diffuse nebulae* (also known as *emission nebulae*), which are the actual birthplaces of the stars. These gossamer, glowing clouds of hydrogen gas line the spiral arms of our Galaxy, lying in or near the galactic plane. Of the hundreds that have been catalogued, the best-known and most striking for stargazers in the Northern Hemisphere is the magnificent Orion Nebula (M42/M43) of the winter sky. This huge, glowing, greenish mass spans more than a degree in size and is visible to the unaided eye as a misty-looking 4th-magnitude star in Orion's sword. Fascinating even in binoculars, it becomes ever-more entrancing as optical aperture increases. At its core lies the famed Trapezium multiple star, resolvable even in a 2-inch glass at 25× and the brightest members of what is actually a star cluster in formation. Three other prominent objects of this class are the Lagoon Nebula (M8) with its embedded star cluster NGC 6530, the Trifid Nebula (M20), and the Horseshoe/Omega/Swan Nebula (M17) – all lying near each other in the constellation Sagittarius. Southern Hemisphere observers have the awesome Tarantula Nebula (NGC 2070), located within the Large Magellanic Cloud in the constellation Dorado. Even though it's in another galaxy and lies at a distance of 170,000 light-years from us, it's clearly visible to the unaided eye and is an impressive sight in binoculars. The view in even a small telescope is breathtaking. If this nebula were as close to us as is the Orion Nebula, it would span more than 30 degrees of the sky and shine with a total brightness three times that of Venus!

A second and quite different-looking type of nebulosity is that of the *planetary nebulae*. While most of the 1,500 examples known lie in or near the Milky Way's misty band, they can also be found scattered all over the sky. Representing the opposite end of stellar evolution from the diffuse nebulae, these fascinating and often eerie-looking balls and rings and shells of gas have been ejected by dying stars in the late stages of their lives. The name was coined by Sir William Herschel after finding a number of these objects during his famous surveys (or "sweeps") of the sky that looked like the planet Uranus, which he had just recently discovered – a small, round, greenish-blue disk of high surface brightness as seen in the eyepiece of his various home-made, metal-mirrored reflecting telescopes. (It's the high surface brightness of planetaries compared with other nebulae that makes it possible to view them through light pollution, muggy skies, and bright moonlight.) Herschel's first find was the Saturn Nebula (NGC 7009) in Aquarius, quickly followed by many others.

Figure 13.7. The Orion Nebula (M42/M43) – considered by many observers to be the finest diffuse nebula in the sky and the most awesome deep-sky wonder of them all. New stars are being born within this vast stellar nursery as we watch! The Trapezium multiple star lies at its core (its image overexposed here), supplying much of the illumination for the cloud itself. Visible to the unaided eye and in binoculars as a fuzzy-looking "star" in the middle of Orion's sword, its aspect in telescopes is simply overpowering. While films see the nebulosity as predominantly reddish, pinkish, and bluish, visually it appears a distinct emerald-turquoise in hue, its embedded stars looking like diamonds on green velvet. M43 is the smaller patch of nebulosity just above and attached to M42, the main nebula itself in this stunning image taken with a 108-mm apochromatic refractor. Courtesy of Steve Peters.

Figure 13.8. Two more of the sky's best diffuse nebulae – the big, bright Lagoon (M8) at bottom and the smaller Trifid (M20) at top – photographed with an 8-inch Newtonian reflector. Lying just 90 arc-minutes (1.5 degrees) apart in Sagittarius, both clouds fit into the field of view of binoculars and RFTs. Note the open cluster NGC 6530 embedded within the Lagoon, from which its stars were born. As with the Orion Nebula, film sees these nebulosities as red, but here (especially in the case of the dimmer Trifid) little color of any kind is visible except in large backyard telescopes. Courtesy of Steve Peters.

But the most famous and prominent of all planetaries are two from the already-existing Messier catalog – the Ring Nebula (M57) in Lyra and the Dumbbell Nebula (M27) in Vulpecula. The latter is so big and bright that it's visible in 7 × 50 binoculars. Some other favorites are Jupiter's Ghost (NGC 3242) in Hydra, the Eight-Burst Nebula (NGC 3132) in Vela, the Eskimo/Clownface Nebula (NGC 2392) in Gemini, and the Cat's Eye/Snail Nebula (NGC 6543) in Draco. These are all within reach of even a 2- or 3-inch glass and are wonderful sights in 6-inch and larger telescopes. The late, beloved *Sky & Telescope* writer Walter Scott Houston – whose "Deep-Sky Wonders" column ran monthly for nearly half a century within its pages – described this class of objects as follows: "Delightful planetary nebulae – ephemeral spheres that shine in pale hues of blue and green, and float amid the golden and pearly star currents of our Galaxy . . . on the foam of the Milky Way like the balloons of our childhood dreams. If you want to stop the world and get off, the lovely planetaries sail by to welcome you home."

A third class of nebulosity is that of the *reflection nebulae*. These are typically very faint glows seen around stars and caused by starlight reflecting off dust (and

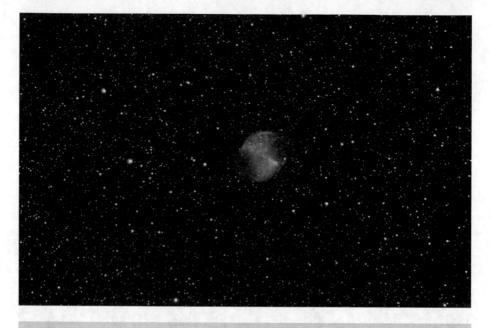

Figure 13.9. This exquisite image of the Dumbbell Nebula (M27) in Vulpecula was made with an 11-inch Schmidt–Cassegrain catadioptric. Generally considered the easiest planetary to see in the entire sky, it can be picked up in finders and binoculars, and is a pretty sight in even the smallest of telescopes. In medium-sized backyard instruments, it looks like a big puffy pillow floating serenely amid the stars of the summer Milky Way. Courtesy of Dennis di Cicco.

to a lesser extent gas) enveloping the stars themselves. One of the more famous of these is Temple's Nebula (NGC 1435), a tear-shaped feeble glow surrounding the star Merope in the Pleiades star cluster in Taurus. Although it was discovered with a 4-inch refractor, it's not an easy object to glimpse. Part of the problem is that this is a winter object, and condensation from the observer's breath onto the cold glass surfaces of eyepieces often tends to put haloes around *all* stars! In the typical high humidity of summer, eyepieces also tend to fog up when the eye is placed near them, resulting in the same effect. This is why many observers (myself included) find the fall with its low humidity and moderate temperatures the best stargazing time of the year. Another noted reflection nebula is the Witch's Head Nebula (IC 2118) in Eridanus, just west of the brilliant star Rigel – itself in Orion and the source of illumination for the nebula's faint glow. More than two degrees across in its longest dimension, it requires a very wide field and a highly transparent night to be seen with certainty. As a class, these objects are visually disappointing and are not apt to be found on anyone's list of personal deep-sky favorites!

A fourth class of visible nebulosity is that of the *supernova remnants*. These are the remains of supernova outbursts within our Galaxy, which are very rare and only a few of which are visible in backyard telescopes. Without question the

Figure 13.10. The spectacular Helix Nebula (NGC 7293) in Aquarius, as photographed with an 11-inch Schmidt–Cassegrain catadioptric. This huge planetary is the death throes of an aged star shedding its outer layers into a celestial smoke-ring. Unlike its famed counterpart, the Ring Nebula in Lyra (M57 – perhaps the best-known of all planetaries), this is not an easy object to see visually. Although it has the highest total apparent brightness of its class (magnitude 6.5), its huge angular size (half that of the Moon) results in an extremely low "surface brightness." Best viewed in low-power, wide-field telescopes (RFTs), it can be glimpsed in binoculars on dark, transparent nights. Courtesy of Steve Peters.

most famous and most easily seen of these is the Crab Nebula (M1) in Taurus. Originally classified as a planetary, this is the expanding cloud of debris ejected by Supernova 1054 AD. Just detectable in 10 × 50 binoculars, a 2-inch glass shows its ghostly glow, and the view becomes ever more fascinating as aperture increases. In large Dobsonian reflectors, the tattered edges of the expanding cloud can be glimpsed, as well as the neutron star/pulsar rapidly spinning (33 times a second!) at its heart. As every deep-sky observer is aware, this was the object that motivated comet-hunter Charles Messier to compile his famed catalogue of "comet impostors." Few today remember his comets, but he has been immortalized through his listing of "nuisances", as he referred to them.

Another supernova remnant that is readily available to the stargazer is the huge Veil/Filamentary/Cirrus/Lacework Nebula (NGC 6960/6992–5) in Cygnus. Here we find two halves of a giant loop (actually a broken bubble) each over a degree in length, spanning nearly 3 degrees of the sky. To be seen in its entirety, rich-field scopes or giant binoculars are required, but each of the individual halves will fit into the field of a low-power, wide-angle eyepiece on normal telescopes. This

Crab Nebula [Messier 1] Supernova remnant
Distance: 6500 light years
Intergrated magnitude: 8.4

Figure 13.11. Lord Rosse's Crab Nebula (M1) in Taurus – the visible remains of the great supernova outburst of 1054 AD, with a rapidly spinning neutron star/pulsar at its core. Readily picked up in a 3-inch telescope as a pale oval glow, the nebula's mottled structure becomes evident in medium-size amateur scopes. However, the filaments and ragged edges seen in this image taken with an 8-inch Schmidt–Cassegrain catadioptric require large backyard instruments to glimpse visually. Note the star lying just above the nebula's center. This is actually a pair of stars (unresolved at this scale), one component of which is the collapsed core of the sun that went supernova. The light from this neutron star flashes 33 times a second as it spins! M1 is the object that led Charles Messier to compile his celebrated catalog of deep-sky objects. Courtesy of Mike Inglis.

object also needs a dark night to be seen, since it is fairly faint and has a low surface brightness.

This is a good opportunity to mention *light-pollution* and *nebula filters*, which are being increasingly used by observers to combat bright skies and enhance the visibility of nebulosities. By isolating certain spectral lines at which these clouds glow, the sky background is suppressed and the nebula appears brighter than in the unfiltered view. The Veil is an excellent target on which to try one of these devices, transforming it from a marginal wonder to a real showpiece of the deep sky. (Note here that nebula filters actually *dim* stars themselves, which do not shine at nebular wavelengths; so objects containing lots of stars – such as star clusters and galaxies –appear fainter as seen through such filters.) In viewing the remnants of exploding stars such as the Crab or Veil, keep in mind that you are seeing, as one writer so well expressed it, "The shards of a supernova; the spawn of minds to come."!

The fifth type of nebulosity does not give off light but rather obscures that of objects shining behind it. These are the *dark nebulae*. Of the thousands that have been cataloged, the most prominent were found long ago by the legendary observer Edward Emerson Barnard and carry the designation "B." The largest and most obvious of these are the huge dark clouds of dust lying between the spiral arms of our Galaxy. These can be seen along many parts of the Milky Way as obvious dark swaths largely vacant of stars on clear moonless nights – one of them is the amazing Great Rift that splits the Galaxy in two in the constellation Cygnus. There are smaller dark clouds such as the 7-degree-long Pipe Nebula (B 59, 65, 66, 67, and 78) and the 30-arc-minute-wide Barnard's S-Nebula (B 72), both in the "dusty" constellation Ophiuchus. These big objects obviously require wide fields of view, being best seen in binoculars and RFTs. But dark nebulae scale all the way down in size to the tiny, speck-like *Bok globules* (which are really condensing, non-luminous proto-stars) found silhouetted against many bright nebulosities and requiring observatory-class telescopes to be seen visually.

Figure 13.12. The famed, often-pictured Horsehead Nebula (Barnard 33/IC 434) in Orion. Although striking photographically (as in this image made with a 108-mm apochromatic refractor), it's among the most difficult of all deep-sky objects to see visually, requiring a dark transparent sky, a well-dark-adapted eye, and averted vision just to glimpse. Experienced observers have detected it with apertures as small as 5- or 6-inch, and report that a nebula filter enhances visibility of the tiny dark Horsehead silhouetted against the background nebulosity. The brightest star seen here is Alnitak (ζ Orionis) in Orion's belt, and to its lower left is the big Flame Nebula (NGC 2024) – a much easier object to see telescopically than is the Horsehead itself. Courtesy of Steve Peters.

The most obvious of the naked-eye dark nebulae is the famed Coal Sack in the far-southern constellation Crux. Measuring 7- by 5-degree across, this great black "hole in the heavens" (a term first coined by Sir William Herschel for a void he found near the globular cluster M80 in Scorpius) is unmistakable. Ironically, it happens to lie right next to one of the brightest and most spectacular star groupings in the entire sky – the glittering Jewel Box Cluster (NGC 4755). But the best-known and most-photographed dark nebula of them all is one that's definitely *not* a naked-eye object! It's the infamous Horsehead Nebula (B 33) in Orion – a tiny, dark protrusion 5-arc-minutes long that's silhouetted against the faint glow of the emission nebula IC 434, which extends south of the star Alnitak (ζ Orionis) in the Hunter's belt. Visually, this is one of the most challenging objects of its kind in the entire heavens. Although it has been glimpsed with apertures as small as 5-inch (and reportedly even in 10×70 binoculars), if you hope to see it you will typically need at least an 8-inch telescope, a nebula filter (to enhance the emission nebulosity, thereby increasing contrast with the Horsehead), a dark transparent night, and a completely dark-adapted eye. While this appears to be everyone's favorite dark nebula (solely from its photographic appearance), it – like the reflection nebulae – is unlikely to be found on many observers' "hit lists." Indeed, few of the showpiece rosters which the author has collected from around the world and from all time periods have ever included either of these types of non-luminous denizens of the deep sky. And with good reason!

Galaxies and Quasars

We now come to the remote *galaxies* themselves. These colossal, spinning star-cities containing perhaps half a *trillion* suns range from large, bright, naked-eye objects to ultra-faint smudges of light at the limits of the world's largest telescopes. Hundreds of billions of them exist within the observable universe, and tens of thousands lie within range of large backyard telescopes. And while relatively few are to be seen near the plane of the Milky Way because they are obscured by its clouds of stars and dust, great swarms of them are found all over the sky away from the Galaxy – especially near the galactic poles, where we're looking out unobscured into deep space. Among the galaxies visible without optical aid (besides our own Milky Way, discussed below), the Great Spiral in Andromeda – or officially, the Andromeda Galaxy (M31) – surely stands without equal for Northern Hemisphere observers. Clearly visible as an elongated glow on a dark night just northwest of the star ν Andromedae, it's easily found by imagining a line from the star in the bottom right-hand corner of the Great Square of Pegasus (another asterism!) to the upper left-hand one and extending it by its own length. It's been traced to more than 5 degrees (or 10 full-Moon diameters!) in extent using binoculars, and its star-like nucleus, nuclear bulge and disk, dust lanes and spiral arms are all visible through amateur telescopes, seen from across a distance of some 2,400,000 light-years. Another marginally visible naked-eye spiral is the Pinwheel or Triangulum Galaxy (M33) in the adjoining constellation Triangulum. Larger than the full Moon in apparent size, it becomes well defined in binoculars and is a fascinating

Figure 13.13. The magnificent Triangulum Galaxy (M33), as imaged with an 11-inch Schmidt–Cassegrain catadioptric. This graceful spiral is a member of the Local Group of galaxies, along with the nearby Andromeda Galaxy and our own Milky Way. Visible in binoculars on dark, moonless nights, it's so big (larger than the apparent size of the full Moon) and it's light so spread out that it's often passed right over when sweeping with telescopes, despite the fact that it shines at magnitude 5.7! (M33 has been seen with the unaided eye by experienced observers.) In 6-inch and larger telescopes, much structure is visible, including its delicate spiral arms. At a distance of some 2,700,000 light-years, it lies only slightly further from us than does its famous neighbor in Andromeda. Courtesy of Dennis di Cicco.

sight telescopically, despite its relatively low surface brightness compared with its famed neighbor.

Southern Hemisphere observers have two marvelous and obvious naked-eye galaxies gracing their sky. These are the Large Magellanic Cloud (LMC) spanning the Dorado–Mensa border and the Small Magellanic Cloud (SMC) in Tucana. Lying within 20 degrees of the South Celestial Pole and just 22 degrees from each other, they appear like large, detached portions of the Milky Way and are quite bright – so much so that the LMC itself can be seen even in full moonlight! Visually, the LMC spans some 6 degrees in extent and the SMC 3.5 degrees. Both are classified as dwarf irregular galaxies and are satellites of our own Milky Way. As such, they are the very closest of all galaxies to us, the LMC being 170,000 and the SMC 200,000 light-years distant. Along with the Milky Way itself, as well as M31 and M33, they are members of the Local Group of galaxies containing some three dozen systems. Even a pair of binoculars will resolve the two Clouds into their individual stars, and telescopes reveal an amazing wonderland of deep-sky objects within them – many having their own individual NGC numbers.

Figure 13.14. Lord Rosse's famed Whirlpool Galaxy (M51) in Canes Venatici, showing its bright nucleus, open spiral arms, and fainter companion galaxy NGC 5195. Seen face-on, in a 3- or 4-inch glass this object looks like two dim, unequally sized nebulae nearly in contact, while the graceful spiral structure becomes visible in 8-inch and larger telescopes on dark, transparent nights. It lies at a distance of some 30,000,000 light-years from us, which is typical of many of the brighter galaxies seen in amateur instruments. This photograph was taken with an 8-inch Newtonian reflector. Courtesy of Steve Peters.

Among other prominent individual galaxies visible in binoculars and small telescopes are the Whirlpool (M51) in Canes Venatici, the Blackeye (M64) in Coma Berenices, and the Sombrero (M104) in Virgo. Like double stars, galaxies often occur in pairs (and larger groupings), some of the more spectacular being M81/M82 in Ursa Major, and M65/M66 and M95/M96 in Leo. And just as stars form clusters, so too do galaxies. We've already mentioned the Local Group, but it is only part of a much larger assemblage known as the Virgo Cluster (sometimes referred to as the Coma–Virgo Cluster) containing thousands of galaxies. This exciting region of the sky is popularly known as the "Realm of the Nebulae" – a name dating back to a time before the galactic nature of these nebulous-looking objects was known. Sweeping across the core of this swarm with a 4-inch telescope reveals dozens of them (many having Messier designations), while 8-inch

Figure 13.15. In striking contrast to the Whirlpool's aspect is this edge-on system known as the Sombrero Galaxy (M104) in Virgo. This eyepiece sketch shows its appearance in a 13-inch refractor, but the nuclear bulge is obvious even in a 3-inch glass and the dark equatorial dust-lane can be glimpsed with a 6-inch using averted vision. The Sombrero is so bright that it can be seen in full moonlight – this, despite its distance of 28,000,000 light-years! South is up, as viewed in an inverting telescope.

and larger instruments show so many galaxies that identification becomes difficult. There are places within the cluster where a wide field of view will contain half-a-dozen or more of these remote star-cities – all softly glowing from across some 55,000,000 light-years of intergalactic space.

Amazingly, the Virgo Cluster is itself part of a much larger structure known as the Local Supercluster containing *tens of thousands* of individual galaxies! Other huge clusters of galaxies are to be found around the sky; three visible in large amateur scopes (14-inch aperture and up) are the Coma, Hercules, and Perseus Galaxy Clusters, lying at distances of 350 to 500 million light-years from us.

But the remotest denizens of deepest space are the famed *quasars* (or quasistellar objects). Now believed to be caused by galactic mergers feeding an immense central black hole, resulting in the release of an incredible amount of energy, these bizarre objects give off as much light as hundreds of galaxies combined, making them visible from across vast distances. The "closest" and brightest of them all – and the first one to be recognized – is known as 3C273 (its designation in the third Cambridge catalogue of radio sources). Appearing like a dim 13th-magnitude bluish star in the constellation Virgo, this object lies nearly 2 *billion* light-years from us – yet it is visible in 4- to 6-inch backyard telescopes! (Incidentally, I published the very first article ever written on looking at quasars with amateur-class instruments, in the December, 1979, issue of *Astronomy* magazine.)

Two final aspects of observing galaxies need mention. One is the visual discovery of supernova outbursts in the brighter nearby spirals by amateur astronomers, some using telescopes as small as a 4-inch glass. And here the world's record holder is Reverend Robert Evans of Australia, who has found some two dozen of them to date using his 10- and 16-inch reflectors. Becoming intimately familiar with the appearance of a selected list of targets, and then checking them night after night, is sure to yield yet more discoveries by those who love staring at galaxies. If you believe you've found a supernova – a "star" that wasn't there before, appearing in or near a galaxy – then, as for comets and ordinary novae, you should report it immediately to the International Astronomical Union's Central Bureau for Astronomical Telegrams, at the Harvard–Smithsonian Center for Astrophysics, 60 Garden Street, Cambridge, Massachusetts 02138 – or preferably by e-mail to cbat@cfa.harvard.edu.

A second important aspect of viewing galaxies is recognition of the fact that you are seeing the most distant objects in creation by means of light millions or even billions of years old. Your telescope is focusing archaic photons onto the retina of your eye that left these regions eons ago! (See the quotations and discussion about the "photon connection" in the next chapter.)

Milky Way

We have yet to mention the largest, brightest, and most magnificent galaxy in the sky – we're speaking, of course, about our very own *Milky Way* galaxy! Indeed, this is surely the finest "deep-sky" object of any type. Since we are embedded in one of its spiral arms and since this colossal starry pinwheel circles the entire visible heavens, the instrument of choice here is the unaided human eye itself, followed by wide-angle binoculars and then telescopes (especially RFTs). Using any of these to survey its massed star-clouds – particularly in regions such as Cygnus, Scutum, and Sagittarius – on dark, transparent nights is truly an exhilarating experience.

With the eye alone, you get the definite feeling of being submersed within the Galaxy's stars as it encompasses you. The brightest parts of the Milky Way are found in the summer sky, where we're looking inward toward its core (whereas in winter we are looking out through a thinner stratum of stars to its edge). Two such areas stand out as superb binocular and telescopic sights. One is the Scutum Star Cloud in the constellation of the same name. Also referred to as the "Gem of the Milky Way" and "Downtown Milky Way," this 12- by 9-degree expanse of the heavens is a binocular and telescopic wonderland. The other area is the Small Sagittarius Star Cloud, a 2- by 1-degree piece of sky near the galactic center in Sagittarius that seemingly holds infinite numbers of starry pinpoints. Slowly sweeping across either of these amazing star-clouds on dark nights with a 4- or 6-inch RFT at 15× to 25× is truly an experience never to be forgotten.

Also available for exploration is the much bigger Large Sagittarius Star Cloud, which looks to the unaided eye like steam escaping from the spout of the well-known Teapot asterism of Sagittarius.

Figure 13.16. Milky Way star-clouds in Cygnus, as photographed with a 300-mm f/2.8 camera centered on the North America Nebula (NGC 7000). In sweeping here with binoculars and RFTs – and realizing that the brighter stars in the cloud are closer to you than are the fainter ones – look for the striking illusion of depth! (This is evident even in this image, where the nebula appears to lie *beyond* the foreground stars.) Spanning nearly two degrees in extent, the nebula itself can be seen in binoculars – and even glimpsed with the unaided eye – on extremely dark, transparent nights. Our Milky Way is without question the grandest galaxy and deep-sky wonder of them all! Courtesy of Steve Peters.

But there's more excitement to share about the Milky Way, and it involves the very striking and surprising illusion of perceiving depth in its structure! In viewing the great billowy star-clouds of rich regions such as Sagittarius, Scutum, and Cygnus with the unaided eye and binoculars, be alert for an amazing 3-D effect that can occur without warning. As the eye–brain combination makes the association that the fainter stars you're seeing lie further away from you than do the brighter ones – *that you're actually looking at layer upon layer of stars* – the Milky Way can suddenly jump right out of the sky at you as the three-dimensional starry pinwheel it really is!

Navigating the Great Beyond

The fact that this is one of the longest chapters in this book indicates both the extent of the field of deep-sky observing and its wide popularity with today's amateur astronomer. And so it is that the recommended references to this activity in what follows are much more extensive than those given in any previous chapter. An immediate and useful one is the Showpiece Roster that appears in Appendix 3. It offers outstanding specimens of some of the finest first-magnitude and highly tinted stars, double and multiple systems, stellar associations and asterisms, open and globular star clusters, diffuse and planetary nebulae, supernova remnants, and galaxies for viewing with backyard telescopes in the 2- to 14-inch aperture range.

But here already there arises the crucial matter of finding your way around the sky to locate these wonders. Even if you are using mechanical or digital setting circles – or the more modern Go-To or GPS technology – to "navigate" your telescope, you will need to know at least where the brighter stars are, if for no other reason than when providing initial alignment for your telescope's computer. And as I have previously indicated, the real fun of leisurely stargazing (at least for many of us purists) is traditional "star-hopping" to targets – which means using a good star atlas. While standbys such as *Norton's Star Atlas*, *The Cambridge Star Atlas*, and the *Bright Star Atlas* are fine for general navigation of the heavens, successful star-hopping to the sky's fainter denizens requires a more detailed celestial roadmap – one showing stars to at least as faint as 8th magnitude, or about the limit of most finders and binoculars. While there are several highly detailed, multi-volume atlases readily available today (such as *Uranometria 2000.0*, which plots over 280,000 stars to magnitude 9.75 and shows more than 30,000 non-stellar objects), an ideal one for deep-sky observers is *Sky Atlas 2000.0*, already referenced in Chapter 12. Its more than 81,000 stars to magnitude 8.5 (plus some 2,700 non-stellar wonders) provide an ideal number of signposts for effective star-hopping to targets when using typical finders or binoculars, without resulting in overkill. Even if the telescope you already own or plan to buy offers automated finding, I urge you to still spend at least some time star-hopping your way around the majestic highways and byways of the night sky. You'll be delighted you did!

Next come several true classics, found today as reprints of early works and/or original editions from various astronomy publishers, in stores selling new and used books, and in many libraries. Two charming nineteenth-century guides to the deep sky for amateurs are *A Cycle of Celestial Objects*, Volume Two, by W.H. Smyth, which is known as *The Bedford Catalogue* (Willman-Bell, 1986), and Volume Two of *Celestial Objects for Common Telescopes* by T.W. Webb (Dover Publications, 1962). Both contain extensive detailed lists and descriptions of deep-sky wonders (over 800 of them in the first case and some 3,000 in the second). Two more recent works are *1001 Celestial Wonders* by C.E. Barns (Pacific Science Press, 1929) and William Tyler Olcott's *Field Book of the Skies* (G.P. Putnam's Sons, 1929 – with additional printings up through the mid-1950s). The latter guide's marvelous system of dual sky maps – one for the naked eye and binoculars (or "field glasses" as Olcott referred to them), the other for telescopes – cover nearly 500 easily found deep-sky wonders, including variable and double stars, visible in its author's 3-inch refractor. So revered was this work that it was often referred to as "the astronomer's *Bible*." An even more recent classic reference is the truly monumental *Burnham's Celestial Handbook* by Robert Burnham, Jr. (Dover Publications, 1978). In three thick volumes totaling 2,138 pages, it provides descriptions of and/or data for more than 7,000 deep-sky objects of all types, plus many fascinating essays on various aspects of astronomy, including its history, star and constellation lore, and the aesthetic/philosophical/spiritual aspects of stargazing itself.

Some of the many modern books and guides devoted to deep-sky observing follow. Note in particular the first two references, which are by the legendary visual observer Stephen James O'Meara. (Among his many amazing feats is being the first person to see Halley's Comet more than a year before its scheduled 1986 return while it was still in the outer Solar System, and the visual detection of the mysterious radial "spokes" in Saturn's rings long before the Voyager spacecraft found them!) Another highlight is the collected works (edited by O'Meara) of famed observer and *Sky & Telescope* columnist Walter Scott Houston. A kind and wise mentor to generations of deep-sky observers worldwide, Scotty (as he was known to many) passed away a few years ago to roam his beloved heavens. And now, the references themselves:

Deep-Sky Companions: The Messier Objects – Stephen James O'Meara (Sky Publishing & Cambridge University Press, 1998)

Deep-Sky Companions: The Caldwell Objects – Stephen James O'Meara (Sky Publishing & Cambridge University Press, 2002)

Deep-Sky Wonders – Walter Scott Houston (Sky Publishing, 1999)

Messier's Nebulae and Star Clusters – Kenneth Glyn Jones (Cambridge University Press, 1991)

The Messier Album – John Mallas and Evered Kreimer (Sky Publishing, 1978)

Star-Hopping for Backyard Astronomers – Alan MacRobert (Sky Publishing, 1993)

Field Guide to the Deep Sky Objects – Mike Inglis (Springer, 2001)

Astronomy of the Milky Way, subtitled (in two volumes) *An Observer's Guide to the Northern Sky* and *An Observer's Guide to the Southern Sky* – Mike Inglis (Springer, 2004)

Observing the Caldwell Objects – David Ratledge (Springer, 2000)

Deep Sky Observing: The Astronomical Tourist – Steven Coe (Springer, 2000)

NGC 2000.0: The Complete New General Catalogue and Index Catalogues of Nebulae and Star Clusters by J.L.E. Dreyer – Roger Sinnott (Sky Publishing, 1988)

Observing Handbook and Catalogue of Deep-Sky Objects – Christian Luginbuhl and Brian Skiff (Cambridge University Press, 1990)

The Night Sky Observer's Guide: Volume 1 Autumn & Winter, Volume 2 Spring & Summer – George Kepple and Glen Sanner (Willmann-Bell, 1998)

DeepMap 600 – Wil Tirion and Steve Gottlieb (Orion Telescopes & Binoculars, 1999)

Celestial Harvest: 300-Plus Showpieces of the Heavens for Telescope Viewing & Contemplation – James Mullaney (Dover Publications, 2002)

Sky & Telescope's Messier Card (Sky Publishing, 2003)

Sky & Telescope's Caldwell Card (Sky Publishing, 2002)

Sky & Telescope's Pocket Sky Atlas – Roger Sinnott (Sky Publishing, 2006)

Mention should be made too of the valuable eight-volume series entitled *Webb Society Deep-Sky Observer's Handbook* covering various classes of objects. They were published between 1979 and 1982 by Enslow Publishers and edited by Kenneth Glyn Jones. Information about these observing guides, as well as about the Society and its journal *The Deep-Sky Observer*, can be obtained at <u>www. webbsociety.freeserve.co.uk</u>.

I must tell you, too, about a most wonderful book entitled *New Handbook of the Heavens* by H.J. Bernhard, D.A Bennett, and H.S. Rice – originally published by McGraw-Hill in several hardcover printings from 1941 to 1956, and subsequently in paperback by New American Library/Mentor Books in 1954 with numerous printings running into the early 1960s. While it is not a deep-sky guide as such, its three chapters on observing double and multiple stars, variable stars, and star clusters and nebulae are without question the finest and most thrilling ever written on these objects. Sadly, it can only be found today on the used-book market – and then only if you're lucky. Finally, as a reminder, references for observing first-magnitude, highly tinted, variable, double, and multiple stars (which, again, *are* themselves deep-sky objects!) were given in Chapter 12.

Perhaps the ultimate test of an observer's ability to navigate the deep sky is that provided by the many *Messier Marathons* that have become so popular with certain observers in recent years. Each spring, it's possible to view all 109 M-objects in a single observing session spanning the night from dusk to dawn! While this may be an interesting challenge to some, it's decidedly *not* the way to enjoy celestial pageantry, for it results in sensory saturation – an overloading of the eye–brain system. As one observer wisely pointed out, all deep-sky objects deserve to be stared at for a full 15 minutes in order to *really see* what you're looking at. Jumping rapidly from one object to another is like reading only the *Cliff's Notes* to the world's great novels. (While I was comparing final showpiece selections for my *Sky & Telescope* series reprint, "The Finest Deep-Sky Objects", first published in 1966, I admit to having viewed as many as 600 objects in a single night using a 13-inch refractor – all found by star-hopping! But a typical evening of leisurely stargazing involves viewing only a dozen or so wonders – especially if the Moon and one or more of the planets happen to be up.)

CHAPTER FOURTEEN

Conclusion

Sharing the Wonder

While stargazing can be a rewarding activity on a strictly personal basis and enjoyed alone, it is in sharing your observations with other people (and organizations, if you're so inclined) that its pursuit realizes its full value. As the old English proverb states: "A joy that's shared is a joy made double." At its simplest level, letting family, friends, neighbors, beginning stargazers – and even total strangers passing by – see celestial wonders through your telescope will bring not only delight to them, but also immense satisfaction to yourself. Just watch the astonishment on their face as they stare at the Moon's alien landscape or the magnificent ice-rings of Saturn or the heavenly hues of a double star like Albireo. Perhaps it will be *you* who first opens up a totally unexpected and awesome new universe to them!

Reporting observations takes you a step beyond this basic (but very important!) level into sharing what you see in the heavens with other stargazers like yourself. This course may be as simple as sending descriptions of selected celestial objects to your local astronomy club newsletter or to one of the multitude of Internet astronomy chat groups. Going even further, you may wish to submit observing notes or even entire articles to the various astronomy magazines such as *Sky & Telescope* (www.skyandtelescope.com), *Astronomy* (www.astronomy.com), and *Astronomy Now* (www.astronomynow.com/magazine.html) for possible publication. There are also the journals of various active amateur groups, including the Webb Society, the British Astronomical Association, the Astronomical League, or the Royal Astronomical Society of Canada. If – and when – the desire to potentially contribute to our knowledge of the universe strikes you

Figure 14.1. A battery of large Dobsonian reflectors manufactured by Obsession Telescopes seen at one of the many local and national star parties that are so popular with stargazers today. Obsession's large "Dobs" are widely considered to be the "Cadillac" of such instruments because of their excellent optics and fine workmanship. Courtesy of Obsession Telescopes.

through participating in such advanced activities as comet seeking, making variable-star magnitude estimates, measuring binary stars, or patrolling the brighter galaxies for possible supernova outbursts, as discussed in Chapters 11, 12, and 13, there are many national and international organizations (both amateur and professional) to which you can submit your observations. A number of the better-known and most active of these are mentioned in those same chapters, along with information for contacting them.

Pleasure Versus Serious Observing

There is rampant in the field of amateur astronomy today a real belief that you must be doing "serious work" of value to science – preferably with sophisticated (and, therefore, expensive) telescopes and accessories – in order to call yourself an observer. This is truly unfortunate, for it has undoubtedly discouraged many budding stargazers from pursuing astronomy as a pastime. The root of the word "amateur" is the Latin word *amare* – which means "to love" – or more precisely, from *amator*, which means "one who loves." An amateur astronomer is one who loves the stars – loves them for the sheer joy of knowing them. He or she may in time come to love them so much that there will be a desire to contribute something to our knowledge of them. But this rarely happens initially. First comes a period of time (for some, a lifetime!) getting to know the sky, and enjoying its treasures and wonders as a celestial sightseer.

Certainly one of the best examples of this is the legendary American observer Leslie Peltier, who developed an early love of the stars. He went on to discover or co-discover 12 comets and six novae, and contributed over 100,000 visual magnitude estimates of variable stars to the American Association of Variable Star Observers. Born on a farm in Ohio where he made his early observations as a boy and young man, he stayed in his rural community and close to nature the remainder of his life – this, despite offers to join the staffs of several major professional observatories! If there is any reader of this book who has not already done so, I

Figure 14.2. A serious (really serious!) observing setup typical of that used by some advanced amateur astronomers today. The main instrument is a 16-inch Cassegrain reflector with a multi-color photoelectric photometer attached. An 8-inch Schmidt–Cassegrain and a 4-inch refractor ride piggyback, equipped for electronic CCD imaging and automatic guiding, respectively. Only one thing seems to be missing here – an eyepiece with which to look through one of the telescopes! Photo by Sharon Mullaney.

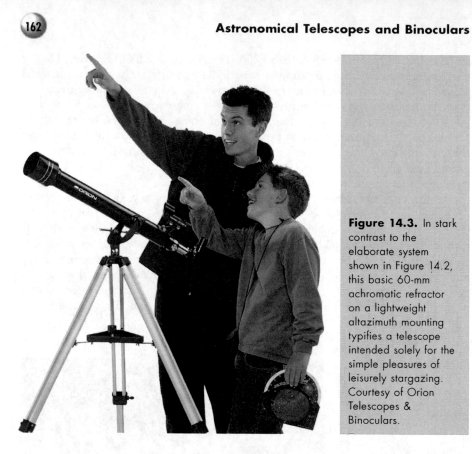

Figure 14.3. In stark contrast to the elaborate system shown in Figure 14.2, this basic 60-mm achromatic refractor on a lightweight altazimuth mounting typifies a telescope intended solely for the simple pleasures of leisurely stargazing. Courtesy of Orion Telescopes & Binoculars.

urge you to obtain and devour a copy of Peltier's autobiography *Starlight Nights* (Sky Publishing, 2000), as an inspiring account of what it truly means to be a "stargazer."

The well-known British amateur James Muirden, in his *The Amateur Astronomer's Handbook* (Harper & Row, 1983), discusses many areas of observational astronomy in which stargazers can become actively involved with programs of potential value to the field. Yet, he wisely advises his readers to "... also never forget that astronomy loses half its meaning for the observer who never lets his telescope range across the remote glories of the sky 'with an uncovered head and humble heart'." He goes on to lament that "The study of the heavens from a purely aesthetic point of view is scorned in this technological age." How very sad!

Aesthetic and Philosophical Considerations

Expanding on the theme of Muirden's comments in the previous section, for many it's the aesthetic and philosophical (and, for some, spiritual) aspects of astronomy that constitute its greatest value to the individual. No other field of human endeavor is so filled with inspiring vistas of radiant beauty, infinite diversity, heavenly-hued pageantry, and elevating heady adventure as is stargazing. And here as in no other field of science we have an opportunity not only to see and experience the universe at its grandest, but to come into actual physical contact with it through the amazing "photon connection."

This is a term that I coined in the June, 1994, issue of *Sky & Telescope* magazine. It involves the fact that the light by which we see celestial objects such as stars and nebulae and galaxies consists of photons, which have a strange, dualistic nature. They behave as if they are *both* particles and waves – or particles moving in waves, if you like. Something that was once inside that object has traveled across the vastness of space and time, and ended its long journey on the retina of your eye. *You are in direct physical contact with the object you are viewing*! Little wonder the poet Sarah Teasdale in looking at the stars said, "I know that I am honored to be witness of so much majesty."

Stargazers themselves have been variously referred to over the years as "Naturalists of the Night," "Harvesters of Starlight," "Time Travelers," "Star Pilgrims," and "Citizens of Heaven." And here we conclude this book by sharing the thoughts of such as these – from observers of the sky past and present, both amateur and professional – for your contemplation. As you let these words penetrate your mind and heart, you'll come to find that path to the heavens that's ideally suited for you in your ongoing cosmic adventure as "one who loves" the stars!

"A telescope is a machine that can change your life." – Richard Berry

"Stargazing is that vehicle of the mind which enables us all to roam the universe in what is surely the next best thing to being there." – William Dodson

"But let's forget the astrophysics and simply enjoy the spectacle." – Walter Scott Houston

"I became an astronomer not to learn the facts about the sky but to feel its majesty." – David Levy

"I am because I observe." – Thaddeus Banachiewicz

"Even if there were no practical application for the serene art of visual observing, it would always be a sublime way to spend a starry night." – Lee Cain

"To me, astronomy means learning about the universe by looking at it." – Daniel Weedman

"Astronomy is a typically monastic activity; it provides food for meditation and strengthens spirituality." – Paul Couteau

"Astronomy has an almost mystical appeal. . . . We should do astronomy because it is beautiful and because it is fun." – John Bahcall

"The great object of all knowledge is to enlarge and purify the soul, to fill the mind with noble contemplations, and to furnish a refined pleasure." – Edward Everett

"The true value of a telescope is how many people have viewed the heavens through it." – John Dobson

"The appeal of stargazing is both intellectual and aesthetic; it combines the thrill of exploration and discovery, the fun of sight-seeing, and the sheer joy of firsthand acquaintance with incredibly wonderful and beautiful things." – Robert Burnham, Jr.

"Whatever happened to what amateur astronomers really care about – simply enjoying the beauty of the night sky?" – Mark Hladik

"I would rather freeze and fight off mosquitoes than play astronomy on a computer." – Ben Funk

"Were I to write out one prescription to help alleviate at least some of the self-made miseries of mankind, it would read like this: 'One gentle dose of starlight to be taken each clear night just before retiring'." – Leslie Peltier

"The high-tech devices pervading the market are ruining the spirit of the real meaning of recreational astronomy – feeling a close, personal encounter with the universe." – Jorge Cerritos

"The sky belongs to all of us. It's glorious and it's free." – Deborah Byrd

"Time spent with 2-billion-year-old photons is potent stuff." – Peter Lord

"We are needy, self-absorbed creatures whose fundamental instincts are for survival and propagation. Any time we can transcend the tyranny of our genes is precious, and the night sky is a portal to this transcendence." – Peter Leschak

"To me, telescope viewing is primarily an aesthetic experience – a private journey in space and time." – Terence Dickinson

"Observing all seems so natural, so real, so obvious. How could it possibly be any other way?" – Jerry Spevak

"As soon as I see a still, dark night developing, my hear starts pounding and I start thinking 'Wow! Another night to get out and search the universe.' The views are so incredibly fantastic!" – Jack Newton

"Take good care of it [your telescope] and it will never cease to offer you many hours of keen enjoyment, and a source of pleasure in the contemplation of the beauties of the firmament that will enrich and ennoble your life." – William Tyler Olcott

"The amateur astronomer has access at all times to the original objects of his study; the masterworks of the heavens belong to him as much as to the great observatories of the world. And there is no privilege like that of being allowed to stand in the presence of the original." – Robert Burnham, Jr.

"But it is to be hoped that some zealous lover of this great display of the glory of the Creator will carry out the author's idea, and study the whole visible heavens from what might be termed a picturesque point of view." – T.W. Webb

"But aren't silent worship and contemplation the very essence of stargazing?" – David Levy

"Adrift in a cosmos whose shores he cannot even imagine, man spends his energies in fighting with his fellow man over issues which a single look through this telescope would show to be utterly inconsequential." – Dedication of Palomar 200-inch Hale Telescope.

"How can a person ever forget the scene, the glory of a thousand stars in a thousand hues . . .?" – Walter Scott Houston

"Seeing through a telescope is 50 percent vision and 50 percent imagination." – Chet Raymo

"A night under the stars . . . rewards the bug bites, the cloudy nights, the next-day fuzzies, and the thousand other frustrations with priceless moments of sublime beauty." – Richard Berry

"All galaxies [and other celestial wonders!] deserve to be stared at for a full fifteen minutes." – Michael Covington

"It is not accident that wherever we point the telescope we see beauty." – R.M. Jones

"You have to really study the image you see in the eyepiece to get all the information coming to you. Taking a peek and looking for the next object is like reading just a few words in a great novel." – George Atamian

"We as astronomers can always retreat from the turbulence around us to our sanctum sanctorum, the sky." – Max Ehrlich

"The night sky remains the best vehicle of escape I know. Simply . . . staring up at a crystal clear sky takes the weight of the world off your shoulders." – Victor Carrano

"What we need is a big telescope in every village and hamlet and some bloke there with that fire in his eyes who can show something of the glory the world sails in." – Graham Loftus

"Someone in every town seems to me owes it to the town to keep one [a telescope!]." – Robert Frost

"The pleasures of amateur astronomy are deeply personal. The feeling of being alone in the universe on a starlit night, cruising on wings of polished glass, flitting in seconds from a point millions of miles away to one billions of light-years distant . . . is euphoric." – Tom Lorenzin

"I'm a professional astronomer who deeply loves his subject, is continually in awe of the beauty of nature [and] like every astronomer I have ever met, I am evangelistic about my subject." – Frank Bash

"Nobody sits out in the cold dome any more [at Palomar – and nearly all other professional observatories today!] – we're getting further and further away from the sky all the time. You just sit in the control room and watch television monitors." – Charles Kowal

"There is something communal and aesthetically rapturous about original archaic photons directly striking the rods and cones in my eyes through lenses and mirrors. . . . These same photons now impinging on my retina left ancient celestial sites millions of years ago." – Randall Wehler

"Some amateur astronomers, it is said, experience the 'rapture of the depths' when observing the Andromeda galaxy." – Sharon Renzulli

"We have enjoyed knowing the stars. We are among the thousands who have found them old friends, to which we can turn time after time for refreshing thoughts and relief from the worries and troubles of every-day life." – Hubert Bernhard, Dorothy Bennett, and Hugh Rice

"How could I convey the mystical love I feel for the universe and my yearning to commune with it? Gazing into the beginning of everything, we are young once again. The child within us is set free." – Ron Evans

"I believe that in looking out at the stars we meet deep psychological and spiritual needs." – Fr. Otto Rushe Piechowski

"Spending a dark hour or two working through the starry deeps to catch faint, far trophies is remarkably steadying for the soul. The rest of the world falls away to an extent only realized upon reentering it, coming back with a head full of distant wonders that most people never imagine." – Alan MacRobert

"The universe seems to demand that we stay in a state of continual astonishment." – K.C. Cole

". . . the spell with which Astronomy binds its devotees: the fascination and the wonder, not to be put into words, of the contemplation and the understanding of the heavens." – G. de Vaucouleurs

"I can never look now at the Milky Way without wondering from which of those banked clouds of stars the emissaries are coming." – Sir Arthur Clarke

"To turn from this increasingly artificial and strangely alien world is to escape from *unreality*. To return to the timeless world of the mountains, the sea, the forest, and the stars is to return to sanity and truth." – Robert Burnham, Jr.

"Lo, the Star-lords are assembling, And the banquet-board is set; We approach with fear and trembling. But we leave them with regret." – C.E. Barns

Telescope Limiting-Magnitude & Resolution

Listed below are limiting-magnitudes and resolution values for a variety of common-sized (**SIZE** in inches) backyard telescopes in use today, ranging from 2- to 14-inch in aperture. (The 2.4-inch entry is the ubiquitous 60 mm refractor, of which there are perhaps more than any other telescope in the world!)

Values for the minimum visual magnitude (**MAG.**) listed here are for single stars and are only very approximate, since experienced keen-eyed observers may see as much as a full magnitude fainter under excellent sky conditions. Companions to visual double stars – especially those in close proximity to a bright primary – are typically much more difficult to see than is a star of the same magnitude placed alone in the eyepiece field. Among the many variables involved are light pollution, sky conditions, optical quality, mirror and lens coatings, eyepiece design, obstructed or unobstructed optical system, color (spectral type) of the star, and even the age of the observer. Only a few representative limiting magnitudes are given here (in increments of increasing aperture), as an indication of what an observer might typically expect to see in various sizes of telescope.

Three different values in arc-seconds are listed for resolution, which are for two stars of equal brightness and of about the sixth magnitude. These figures differ significantly for brighter, fainter and, especially, unequal pairs. **DAWES** is the value based on Dawes' Limit ($R = 4.56/A$), **RAYLEIGH** on the Rayleigh Criterion ($R = 5.5/D$), and **MARKOWITZ** on Markowitz's Limit ($R = 6.0/D$). Note that in these equations "A" (for aperture) and "D" (for diameter) are the same thing.

SIZE	MAG.	DAWES	RAYLEIGH	MARKOWITZ
2.0	10.3	2.28	2.75	3.00
2.4		1.90	2.29	2.50
3.0	11.2	1.52	1.83	2.00
3.5		1.30	1.57	1.71
4.0	11.8	1.14	1.38	1.50
4.5		1.01	1.22	1.33
5.0		0.91	1.10	1.20
6.0	12.7	0.76	0.92	1.00
7.0		0.65	0.79	0.86
8.0	13.3	0.57	0.69	0.75
10.0	13.8	0.46	0.55	0.60
11.0		0.42	0.50	0.55
12.0		0.38	0.46	0.50
12.5	14.3	0.36	0.44	0.48
13.0		0.35	0.42	0.46
14.0	14.5	0.33	0.39	0.43

Constellation Names and Abbreviations

The following table gives the standard International Astronomical Union (IAU) three-letter abbreviations for the 88 officially recognized constellations, together with both their full names and their genitive (possessive) cases, and order of size in terms of number of square degrees of sky.

ABBREV.	NAME	GENITIVE	SIZE
AND	Andromeda	Andromedae	19
ANT	Antlia	Antliae	62
APS	Apus	Apodis	67
AQR	Aquarius	Aquarii	10
AQL	Aquila	Aquilae	22
ARA	Ara	Arae	63
ARI	Aries	Arietis	39
AUR	Auriga	Aurigae	21
BOO	Bootes	Bootis	13
CAE	Caelum	Caeli	81
CAM	Camelopardalis	Camelopardalis	18
CNC	Cancer	Cancri	31
CVN	Canes Venatici	Canum Venaticorum	38
CMA	Canis Major	Canis Majoris	43
CMI	Canis Minor	Canis Minoris	71
CAP	Capricornus	Capricorni	40
CAR	Carina	Carinae	34
CAS	Cassiopeia	Cassiopeiae	25
CEN	Centaurus	Centauri	9

ABBREV.	NAME	GENITIVE	SIZE
CEP	Cepheus	Cephei	27
CET	Cetus	Ceti	4
CHA	Chamaeleon	Chamaeleontis	79
CIR	Circinus	Circini	85
COL	Columba	Columbae	54
COM	Coma Berenices	Comae Berenices	42
CRA	Corona Australis	Coronae Australis	80
CRB	Corona Borealis	Coronae Borealis	73
CRV	Corvus	Corvi	70
CRT	Crater	Crateris	53
CRU	Crux	Crucis	88
CYG	Cygnus	Cygni	16
DEL	Delphinus	Delphini	69
DOR	Dorado	Doradus	7
DRA	Draco	Draconis	8
EQU	Equuleus	Equulei	87
ERI	Eridanus	Eridani	6
FOR	Fornax	Fornacis	41
GEM	Gemini	Geminorum	30
GRU	Grus	Gruis	45
HER	Hercules	Herculis	5
HOR	Horologium	Horologii	58
HYA	Hydra	Hydrae	1
HYI	Hydrus	Hydri	61
IND	Indus	Indi	49
LAC	Lacerta	Lacertae	68
LEO	Leo	Leonis	12
LMI	Leo Minor	Leonis Minoris	64
LEP	Lepus	Leporis	51
LIB	Libra	Librae	29
LUP	Lupus	Lupi	46
LYN	Lynx	Lyncis	28
LYR	Lyra	Lyrae	52
MEN	Mensa	Mensae	75
MIC	Microscopium	Microscopii	66
MON	Monoceros	Monocerotis	35
MUS	Musca	Muscae	77
NOR	Norma	Normae	74
OCT	Octans	Octantis	50
OPH	Ophiuchus	Ophiuchi	11
ORI	Orion	Orionis	26
PAV	Pavo	Pavonis	44
PEG	Pegasus	Pegasi	7
PER	Perseus	Persei	24
PHE	Phoenix	Phoenicis	37
PIC	Pictor	Pictoris	59
PSC	Pisces	Piscium	14
PSA	Piscis Austrinus	Piscis Austrini	60
PUP	Puppis	Puppis	20
PYX	Pyxis	Pyxidis	65

ABBREV.	NAME	GENITIVE	SIZE
RET	Reticulum	Reticuli	82
SGE	Sagitta	Sagittae	86
SGR	Sagittarius	Sagittarii	15
SCO	Scorpius	Scorpii	33
SCL	Sculptor	Sculptoris	36
SCT	Scutum	Scuti	84
SER	Serpens	Serpentis	23
SEX	Sextans	Sextantis	47
TAU	Taurus	Tauri	17
TEL	Telescopium	Telescopii	57
TRI	Triangulum	Trianguli	78
TRA	Triangulum Australe	Trianguli Australis	83
TUC	Tucana	Tucanae	48
UMA	Ursa Major	Ursae Majoris	3
UMI	Ursa Minor	Ursae Minoris	56
VEL	Vela	Velorum	32
VIR	Virgo	Virginis	2
VOL	Volans	Volantis	76
VUL	Vulpecula	Vulpeculae	55

Celestial
Showpiece Roster

Below are 300 of the finest deep-sky treasures for viewing and exploration with telescopes from 2- to 14-inch in aperture. Nearly all of them can be seen in the smallest of glasses, and many even in binoculars. Arranged in alphabetical order by constellation (which makes it more convenient to pick out objects for a given night's observations than with one ordered by coordinates), it features brief descriptions of each entry. Primary data sources were *Sky Catalogue 2000.0* and the *Washington Double Star Catalog*. Constellation (**CON**) abbreviations are the official three-letter designations adopted by the International Astronomical Union (see Appendix 2.)

Right Ascension (**RA**) in hours and minutes, and Declination (**DEC**) in degrees and minutes, are given for the current standard Epoch 2000.0. Other headings are the class or type of object (**TYPE**),* apparent visual magnitude/s (**MAG/S**) and angular size or separation (**SIZE/SEP**) in arc-seconds. (Position angles for double stars are not given, owing to the confusion resulting from the common use of star diagonals with refracting and compound telescopes, producing mirror-reversed images of the sky. Observers desiring the latest values of these as well as component separations should consult the US Naval Observatory's *Washington Double Star Catalog* on-line at http://ad.usno.navy.mil/wds.) Approximate distance in light-years (**LY**) is also given in many cases.

Double and multiple stars dominate this roster because of their great profusion in the sky and also their easy visibility on all but the worst of nights. This list extends down to −45 degrees Declination, covering that three-quarters of the entire heavens visible from mid-northern latitudes. (Two "must see" showpieces actually lie slightly below this limit.)

*Key: SS = First-magnitude/Highly tinted and/or Variable single star, DS = Double or multiple star, AS = Association or asterism, OC = Open cluster, GC = Globular cluster, DN = Diffuse nebula, PN = Planetary nebula, SR = Supernova remnant, GX = Galaxy.

OBJECT/CON	RA	DEC	TYPE	MAG/S	SIZE/SEP	REMARKS
γ AND	02 04	+42 20	DS	2.3, 5.5	10"	Almach. Brilliant topaz-orange & aquamarine double – superb contrast! B is close blue & green, 61-year binary for 8" & larger scopes. 300LY
59 AND	02 11	+39 02	DS	6.1, 6.8	17"	Neatly matched, easy bluish-white pair.
56 AND	01 56	+37 15	DS	5.7, 5.9	190"	Wide golden pair parked on SW edge of cluster NGC 752. 360LY
NGC 752 AND	01 58	+37 50	OC	5.7	50'	Large, sprawling clan of over 60 stars. 1,200LY
M31/M32/M110 AND	00 43	+41 16	GX	3.5/8.2/8.0	178' × 63'/8' × 6'/17' × 10'	Andromeda Galaxy & companions – magnificent! Nucleus, disk, dust lanes, spiral arms all visible. Binocular wonder! 2,400,000LY
NGC 7662 AND	23 26	+42 33	PN	8.5	32" × 28"	Blue Snowball. Small but striking soft-blue cosmic egg. 5,600LY
NGC 891 AND	02 23	+42 42	GX	10.0	11' × 2'	Often-pictured but dim edge-on galaxy with dust lane. 13,000,000LY
U ANT	10 35	–39 34	SS	5.4-6.8	–	Striking red "carbon" star – seldom observed owing to low altitude.
ζ AQR	22 29	–00 01	DS	4.4, 4.5	2"	Matched, bright, off-white close pair. Famous 850-year binary. 76LY
94 AQR	23 19	–13 28	DS	5.3, 7.3	13"	Lovely pale rose or reddish & light emerald-green double.
M2 AQR	21 34	–00 49	GC	6.5	13'	Stellar beehive – a starburst in larger scopes. 37,000LY
NGC 7009 AQR	21 04	–11 22	PN	8.3	25" × 17"	Saturn Nebula. Striking bright, bluish-green ellipsoid. 3,000LY
15 AQL	19 05	–04 02	DS	5.5, 7.2	38"	Easy, wide duo. Yellowish-orange & ruddy-purple or lilac.
57 AQL	19 55	–08 14	DS	5.8, 6.5	36"	Another roomy pair. Both stars bluish-white – hint of other hues.
V AQL	19 04	–05 41	SS	6.6-8.4	–	A lovely glowing red ember!
γ ARI	01 54	+19 18	DS	4.8, 4.8	8"	Mesarthim. Stunning, perfectly matched blue-white pair! 200LY
λ ARI	01 58	+23 36	DS	4.9, 7.7	37"	Wide color/magnitude-contrast double. 105LY

Name	RA	Dec	Type	Mag.	Sep.	Description
α AUR	05 17	+46 00	SS	0.08	–	Capella. A radiant golden-yellow sun! 42LY
θ AUR	06 00	+37 13	DS	2.6, 7.1	4"	Tight mag.-contrast pair for steady nights. Lilac & yellow. 110LY
14 AUR	05 15	+32 41	DS	5.1, 7.4–7.9	15"	Neat double with variable companion.
UU AUR	06 36	+38 27	SS	5.3–6.5	–	Beautiful red color – a celestial stoplight!
M36 AUR	05 36	+34 08	OC	6.0	12'	Lovely cluster of 60-some stars. 4,000LY
M37 AUR	05 52	+32 33	OC	5.6	24'	Very rich & uniform stellar jewelbox – superb! Best in AUR. 4,500LY
M38 AUR	05 29	+35 50	OC	6.4	21'	A hundred suns arranged in an oblique-cross formation. 4,000LY
α BOO	14 16	+19 11	SS	–0.04	–	Arcturus. A splendid yellowish-orange stellar gem! 37LY
ε BOO	14 45	+27 04	DS	2.5, 4.9	3"	Izar. Bright, tight double – superb pale-orange & sea-green! Struve's "Pulcherrima" (the most beautiful one). Needs good seeing. 160LY
ξ BOO	14 51	+19 06	DS	4.7, 7.0	6"	Striking – yellow & reddish-orange or purple. 150-yr.binary. 22LY
μ BOO	15 24	+37 23	DS	4.3, 7.0, 7.6	108", 2"	Neat triple system! B-C is 260-yr. binary. Yellow, two oranges. 95LY
κ BOO	14 14	+51 47	DS	4.6, 6.6	13"	Pretty double – tints real but elusive.
π BOO	14 41	+16 25	DS	4.9, 5.8	6"	Closer version of κ BOO.
ζ BOO	14 41	+13 44	DS	4.5, 4.6	0.7"	Matched white, ultra-close 125-yr. binary. Stellar egg in small glass.
Struve 1835 BOO	14 23	+08 27	DS	5.1, 7.4	6"	Sweet pair – white & bluish or lilac.
32 CAM	12 49	+83 25	DS	5.3, 5.8	22"	Nice matched off-white pair. Little-known – a pity! 495LY
U CAM	03 42	+62 39	SS	8.1–8.6	–	One of the reddest stars in the sky.
ST CAM	04 51	+68 10	SS	7.0–8.4	–	Another ruddy stellar gem.
NGC 2403 CAM	07 37	+65 36	GX	8.4	18' × 11'	One of the brightest galaxies & finest spirals in sky. 12,000,000LY
ζ CNC	08 12	+17 39	DS	5.6, 6.0, 6.2	0.9", 6"	Close, matched trio with 60-& 1150-yr. periods. All yellow. 70LY
ι CNC	08 47	+28 46	DS	4.2, 6.6	30"	Albireo of Spring. Superb orange & blue pair! 165LY
X CNC	08 55	+17 14	SS	5.7.5	–	A stellar ruby! Tint obvious even in small glass.

OBJECT/CON	RA	DEC	TYPE	MAG/S	SIZE/SEP	REMARKS
M44 CNC	08 40	+19 59	OC	3.1	90'	Beehive Cluster. A sprawling commune of over 50 suns. Best seen in binoculars & wide-field telescopes. 590LY
M67 CNC	08 50	+11 49	OC	6.9	30'	Lovely but overlooked cluster in shadow of Beehive. 2,500LY
α CVN	12 56	+38 19	DS	2.9, 5.5	20"	Cor Caroli. Magnificent blue-white double – one of the finest! 130LY
Y CVN	12 45	+45 26	SS	5.5–6.0	–	La Superba. A fiery reddish-orange interstellar beacon. 400LY
M3 CVN	13 42	+28 23	GC	6.4	16'	Spring's Globular. First bright GC of season – radiant starball! 35,000LY
M51 CVN	13 30	+47 12	GX	8.4	11' × 8'	Rosse's Whirlpool Galaxy. Big, beautiful face-on spiral. 31,000,000LY
M63 CVN	13 16	+42 02	GX	8.6	12' × 8'	Sunflower Galaxy. Like some vast celestial flower. 35,000,000LY
M94 CVN	12 51	+41 07	GX	8.2	11' × 9'	Small, bright tightly-wound spiral. 22,000,000LY
M106 CVN	12 19	+47 18	GX	8.3	18' × 8'	Big, bright & bold spiral for small glasses. 33,000,000LY
NGC 4631 CVN	12 42	+32 32	GX	9.3	15' × 3'	Humpback Whale Galaxy. Large edge-on spiral. 39,000,000LY
α CMA	06 4	-16 43	DS	-1.46, 8.5	7"	Sirius. Blazing blue-white sapphire with famed white-dwarf companion! A 50-year binary, now widening. Just 9LY away!
ε CMA	06 59	-28 58	DS	1.5, 7.5	7"	Adhara. A miniature Sirius – and much easier! 490LY
h3945 CMA	07 17	-23 19	DS	4.8, 6.8	27"	Albireo of Winter. Splendid reddish-orange & greenish-blue pair!
W CMA	07 08	-11 55	SS	6.4–7.9	–	Red ember in nice contrast with surrounding blue-white field stars.
M41 CMA	06 46	-20 45	OC	4.5	38'	Lovely big, bright sparkling clan of 80 suns below Sirius! 2,400LY
τ CMA/NGC 2362	07 19	-24 57	OC	4.1	8'	Tau Canis Majoris Cluster. Small glittering jewelbox of 60 diamonds surrounding a bright central star. 5,400LY

Object	RA	Dec	Type	Mag	Size	Description
α-2/1 CAP	20 18	-12 33	DS	3.6, 4.2	380"	Algiedi. Naked-eye/binocular orange pair with faint comps. at 7" & 46" forming weak double-double. Stars unrelated: 110LY & 700LY!
β CAP	20 21	-14 47	DS	3.4, 6.2	205"	Wide binocular combo – yellowish-orange & sky-blue. 560LY
o CAP	20 30	-18 35	DS	6.1, 6.6	22"	Neat, closely matched blue-white pair for small scopes. Lovely warm-hued gem.
RT CAP	20 17	-21 19	SS	6.5-8.1	–	
M30 CAP	21 40	-23 11	GC	7.5	11'	Pale-white starry globe nicely contrasted with 8th-mag. star. 40,000LY
η CAS	00 49	+57 49	DS	3.4, 7.5	13"	Easter Egg Double. Beautiful yellow & ruddy-purple or garnet color & magnitude-contrast combo. A 480-year binary. Nearby – just 19LY
ι CAS	02 29	+67 24	DS	4.6, 6.9, 8.4	2.5", 7"	Elegant but tight triple system. Hues yellow, lilac & blue. 160LY.
σ CAS	23 59	+55 45	DS	5.0, 7.1	3"	Tight pair with intense bluish & greenish tints. Quite distant – 1,400LY.
Struve 163 CAS	01 51	+64 51	DS	6.8, 8.8	35"	Colorful, unequal faintish pair – ruddy-orange & blue.
Struve 3053 CAS	00 03	+66 06	DS	5.9, 7.3	15"	Beautiful miniature of Albireo in CYG. Yellowish-orange & blue.
M52 CAS	23 24	+61 35	OC	6.9	13'	Rich, triangular-shaped sparkling group of at least 100 stars. 4,000LY
M103 CAS	01 33	+60 42	OC	7.4	6'	A small fan-shaped clan of several dozen suns. 8,000LY
φ CAS/NGC 457	01 19	+58 20	OC	6.4	13'	Owl/ET Cluster. Distinctive splash of 80 suns & two "eyes"! 9,300LY
NGC 7789 CAS	23 57	+56 44	OC	6.7	16'	Caroline Herschel's Cluster. Rich uniform assemblage of more than 300 faint stars against stardust. Wondrous sight on dark night! 6,000LY
ω CEN	13 27	-47 29	GC	3.6	36'	Omega Centauri Cluster. Colossal stellar beehive containing more than a million suns – an amazing spectacle in any size scope! 17,000LY
NGC 5128 CEN	13 26	-43 01	GX	7.0	18' × 14'	Black Belt Galaxy. Large globe split by dark dust lane. 22,500,000LY

OBJECT/CON	RA	DEC	TYPE	MAG/S	SIZE/SEP	REMARKS
β CEP	21 29	+70 34	DS	3.2, 7.9	13"	Neat unequal pair – greenish-white & blue or purple. Exquisite! 980LY
δ CEP	22 29	+58 25	DS	3.5–4.4, 6.3	41"	Striking pale orange & blue gems. Primary prototype of famed Cepheid variables – period 5.4 days. 1,000LY
ξ CEP	22 04	+64 38	DS	4.4, 6.5	8"	Neat bright pair with subtle colors – bluish & yellowish. 80LY
μ CEP	21 44	+58 47	SS	3.4–5.1	–	Herschel's Garnet Star. Reddest naked-eye star in N. sky. 2,800LY
Struve 2816 CEP	21 39	+57 29	DS	5.6, 7.7, 7.8	12", 20"	Striking triple system with double Struve 2819 (7.5, 8.5, 12") in field!
Struve 2840 CEP	21 52	+55 48	DS	5.5, 7.3	18"	Lovely pair – greenish-white & bluish-white.
NGC 40 CEP	00 13	+72 32	PN	10.2	60" × 40"	Dull reddish-grey disk with central star. 3,000LY
NGC 7023 CEP	21 02	+68 12	DN	6.8	18'	Iris Nebula. Bright reflection nebula surrounding 7th-mag. blue star.
NGC 6939/6946 CEP	20 31	+60 38	OC/GX	7.8, 8.9	8'/11' × 10'	Unique cluster-galaxy combo set 38' apart! 4,000LY & 10,000,000LY
γ CET	02 43	–03 14	DS	3.5, 6.2	3"	Close, bright pair with delicate tints – yellow & ashen. 63LY
M77 CET	02 43	–00 01	GX	8.8	7' × 6'	Intense star-like core surrounded by circular haze. 82,000,000LY
24 COM	12 35	+18 23	DS	5.2, 6.7	20"	Vivid orange & blue-green duo – a lovely jewel! 300LY
M53 COM	13 13	+18 10	GC	7.7	13'	A dim ball of minute stars. Needs aperture to really enjoy. 65,000LY
M64 COM	12 57	+21 41	GX	8.5	9' × 5'	Blackeye Galaxy. Superb bright spiral with dark "eye". "Like a colossal pendent abalone pearl in rayless void"! 25,000,000LY
M88 COM	12 32	+14 25	GX	9.5	7' × 4'	Like a miniature Andromeda Galaxy. Stellar nucleus. 40,000,000LY
M99 COM	12 19	+14 25	GX	9.8	5' × 5'	Pinwheel Nebula. Wonderful face-on spiral. 50,000,000LY
NGC 4565 COM	12 36	+25 59	GX	9.6	16' × 3'	Ghostly edge-on spiral with dark equatorial dust lane. 20,000,000LY

Object	RA	Dec	Type	Mag	Sep	Description
MEL 111 COM	12 25	+26 00	OC	1.8	275'	Coma Star Cluster. Large hazy, naked-eye & binocular wonder. 270LY
γ CRA	19 06	-37 04	DS	4.8, 5.1	1.3"	Twin yellowish binary – stars appear in contact. 69LY
ζ CRB	15 39	+36 38	DS	5.1, 6.0	6"	Pretty pair of bluish-white & greenish-white suns.
σ CRB	16 15	+33 52	DS	5.6, 6.6	7"	Like ζ but stars are yellowish. Binary with 1000-yr. period.
δ CRV	12 30	-16 31	DS	3.0, 8.4	24"	Algorab. Nice color & mag. contrast – yellow & violet or lilac. 125LY
Struve 1669 CRV	12 41	-13 01	DS	6.0, 6.1	5"	Neatly-matched close pair of yellowish-white suns.
NGC 4361 CRV	12 24	-18 48	PN	10.3	80"	Large, round dimly glowing nebulous disk. 2,600LY
NGC 4038/4039 CRV	12 02	-18 52	GX	10.7	3' × 2'	Antennae/Ring-Tail Galaxy. Colliding pair of galaxies! 90,000,000LY
α CYG	20 41	+45 17	SS	1.25	–	Deneb. Colossal blue supergiant 60,000 × Sun's brightness! 1,600LY
β CYG	19 31	+27 58	DS	3.1, 5.1	34"	Albireo. One of grandest sights in the heavens! Magnificent topaz & sapphire-blue pair in radiant MW setting. Finest double star. 380LY
o-1 CYG	20 14	+46 44	DS	3.8, 7.7, 4.8	107", 338"	Lovely wide trio – orange, blue & white in rich MW setting. 200LY
δ CYG	19 45	+45 08	DS	2.9, 6.3	2.5"	Bright, close unequal pair – tough but pretty. Greenish-white & ashen. Best seen in larger apertures. An 800-yr.-period binary. 270LY
16 CYG	19 42	+50 32	DS	6.0, 6.1	39"	Lovely matched golden duo in wide field with Blinking Planetary.
61 CYG	21 07	+38 45	DS	5.2, 6.0	30"	Beautiful easy orange pair. Famous as first star to have its distance (parallax) directly measured – 11LYs. Slow 650-year binary.
V460 CYG	21 42	+35 31	SS	5.6-7.0	–	Striking red gem – an unresolved binary harboring a black hole!
M39 CYG	21 32	+48 26	OC	4.6	32'	Large triangular-shaped splash of 30 stars – best in binoculars & RFTs. 890LY
NGC 6826 CYG	19 45	+50 31	PN	8.9	27"	Blinking Planetary. Pale blue disk with obvious 10th-mag. central star. Alternating between direct and averted vision makes it blink! 3,300LY

OBJECT/CON	RA	DEC	TYPE	MAG/S	SIZE/SEP	REMARKS
NGC 6819 CYG	19 41	+40 11	OC	7.3	5'	Foxhead Cluster. Small, dim but rich clan of 150 stars. 7,300LY
NGC 6960/6992-5 CYG	20 51	+31 13	SR	–	70' × 6'/60' × 8'	Veil/Filamentary/Cirrus Nebula. Large ghostly arcs 3 degrees apart from supernova explosion some 5,000 years ago. Best seen in large binoculars & wide-field telescopes. 1,500LY
NGC 7027 CYG	21 07	+42 14	PN	9.0	18" × 11"	Stephan's/Webb's Proto-Planetary. Small, blue & intense! 3,000LY
γ DEL	20 47	+16 07	DS	4.5, 5.5	10"	Stunning golden-yellow & greenish-blue combo – splendid object! "Ghost Double" Struve 2725 (7.6, 8.4, 6") in field. 100LY
μ DRA	17 05	+54 28	DS	5.7, 5.7	2"	Cozy, yellowish-white identical-twin 480-year binary. 82LY
ν DRA	17 32	+55 11	DS	4.9, 4.9	62"	Another pair of perfectly-matched suns, but brighter & much wider than μ. Both white – superb! Nice binocular pair. 120LY
ψ DRA	17 42	+72 09	DS	4.9, 6.1	30"	Pretty yellow & lilac combo – easy for small glass.
17/16 DRA	16 36	+52 55	DS	5.4, 6.4, 5.5	3", 90"	Nice triple system like μ BOO but primary has comp. All white.
41/40 DRA	18 00	+80 00	DS	5.7, 6.1	19"	Pale-yellow pair with 7.5-magnitude star nearby.
RY DRA	12 56	+66 00	SS	6.8–7.3	–	A glowing stellar ruby!
UX DRA	19 22	+76 34	SS	5.9–7.1	–	Another stunning red sun.
NGC 6543 DRA	17 59	+66 38	PN	8.8	22" × 16"	Cat's Eye/Snail Nebula. Bright blue-green egg with 10th-mag. nuclear sun. One of the finest of its class & always above horizon! 3,500LY
NGC 5907 DRA	15 16	+56 19	GX	10.4	12' × 2'	Splinter Galaxy. Long, narrow & dim edge-on spiral. 35,000,000LY
ε = 1 EQU	20 59	+04 18	DS	5.4, 7.1	10"	Neat pair – both yellowish. Primary close (0.7") visual binary. 200LY
λ = 2 EQU	21 02	+07 11	DS	7.4, 7.4	3"	Tight but striking identical-twin suns.
θ ERI	02 58	–40 18	DS	3.2, 4.3	8"	Radiant white, far-south gem! 120LY

Object	RA	Dec	Type	Mag	Size/Sep	Description
32 ERI	03 54	−02 57	DS	4.8, 6.1	7"	Lovely topaz-yellow & sea-green in superb contrast – a beauty! 300LY
o-2 ERI	04 15	−07 39	DS	4.4, 9.5, 11.2	83", 8"	Faint pair an amazing white-dwarf & red-dwarf 248-year binary. 16LY
NGC 1535 ERI	04 14	−12 44	PN	9.4	20" × 17"	Lassell's Most Extraordinary Object. Blue-green "celestial jellyfish."
NGC 1316 FOR	03 23	−37 12	GX	8.8	7' × 6'	Fornax A. Luminous leader of Fornax Galaxy Cluster. 55,000,000LY
NGC 1360 FOR	03 33	−25 51	PN	9.4	6' × 4'	Bright egg-shaped overlooked jewel. 980LY
NGC 1365 FOR	03 34	−36 08	GX	9.5	10' × 6'	One of the finest barred-spirals in the sky.
α GEM	07 35	+31 53	DS	1.9, 2.9, 8.9	4", 72"	Castor. Dazzling blue-white 470-yr. binary – a magnificent sight! Orange comp. is eclipser YY GEM, ranging from 8.9 to 9.6 over 20 hours. A & B are spectroscopic binaries – a vast six-sun system! 52LY
δ GEM	07 20	+21 59	DS	3.5, 8.2	6"	Close yellow & reddish-purple duo. Binary – 1200-year period. 53LY
20 GEM	06 32	+17 47	DS	6.3, 6.9	20"	Neat yellowish-white and bluish-white pair. 450LY
M35/NGC 2158 GEM	06 09	+24 20	OC/OC	5.1/11	28/5'	Lassell's Delight. Big splashy & spectacular stellar jewelbox with tiny remote clan shining dimly on outskirts. Clusters lie at vastly different distances from each other – 2,700LY & 16,000LY!
NGC 2392 GEM	07 29	+20 55	PN	8.3	20"	Eskimo/Clown Face Nebula. Vivid blue disk with 10th-mag. central sun looking like a hazy star at low power. 3,000LY
α HER	17 15	+14 23	DS	3.1-3.9, 5.4	5"	Rasalgethi. Bright, intensely tinted orange & blue-green pair – superb! Primary huge pulsating semi-regular variable – a supersun! 380LY
δ HER	17 15	+24 50	DS	3.1, 8.7	14"	Famed, very delicate optical (unrelated) pair. White & violet. 94LY
κ HER	16 08	+17 03	DS	5.3, 6.5	28"	Striking yellow & garnet jewels!
ζ HER	16 41	+31 36	DS	2.9, 5.5	0.7"	Herschel's Rapid Binary. 34-yr. period – over 6 orbits since discovery! 30LY

OBJECT/CON	RA	DEC	TYPE	MAG/S	SIZE/SEP	REMARKS
ρ HER	17 24	+37 09	DS	4.6, 5.6	4"	Bright, cozy bluish & greenish pair – stunning.
95 HER	18 02	+21 36	DS	5.0, 5.1	6"	Lovely twin suns – amazing "apple-green & cherry-red" tints! 380LY
100 HER	18 08	+26 06	DS	5.9, 6.0	14"	Another matched pair but wider & pale off-white hues. Little-known.
M13 HER	16 42	+36 28	GC	5.9	17'	Hercules Cluster. A magnificent stellar beehive! Fuzz-ball as seen in binoculars, resolved to its glittering core in 6-inch glass. 24,000LY
M92 HER	17 17	+43 09	GC	6.5	11'	Overshadowed Globular. Eclipsed by M13. Intense core. 26,000LY
NGC 6210 HER	16 44	+23 49	PN	9.3	20" × 16"	Small featureless blue disk – needs magnification to enjoy. 3,600LY
NGC 6229 HER	16 47	+47 32	GC	9.4	4'	"Sea-green in starry triangle." Long mistaken for a PN. 90,000LY
ε HYA	08 47	–06 25	DS	3.3, 6.8	3"	Tight 890-yr. binary. Primary is also a 15-yr. visual binary! 150LY
N HYA = 17 CRT	11 32	–29 16	DS	5.8, 5.9	9"	Perfectly matched, yellowish-white twin suns.
54 HYA	14 46	–25 27	DS	5.1, 7.1	9"	Pretty pair for small glass – yellowish & violet tints.
U HYA	10 38	–13 23	SS	4.8–6.5	–	A fiery, reddish-orange stellar gem.
M48 HYA	08 14	–05 48	OC	5.8	30'	Big, bright splendid splash of some 50 stars the size of Moon. 1,900LY
M68 HYA	12 40	–26 45	GC	8.2	12'	Neglected owing to low DEC – needs dark, steady night. 45,000LY
M83 HYA	13 37	–29 52	GX	8.0	11' × 10'	Big, bold face-on spiral – one of brightest in the sky. 10,000,000LY
NGC 3242 HYA	10 25	–18 38	PN	8.6	40" × 35"	Jupiter's Ghost. Superb bright planetary with pale-blue disk as big in apparent size as Jupiter. Also known as the Eye & CBS Neb. 3,300LY
8 LAC	22 36	+39 38	DS	5.7, 6.5, 10.5, 9.3	22", 49", 82"	Blue-white duo – fainter companions form delicate quadruple. 1,900LY
NGC 7243 LAC	22 15	+49 53	OC	6.4	21'	Nice loose clan of 40 stellar gems. 2,800LY

Object	RA	Dec	Type	Mag	Size/Sep	Description
α LEO	10 08	+11 58	DS	1.4, 7.7	177"	Regulus/Indigo Star. Wide mag.-contrast pair with blue-white primary & comp. that's "seemingly steeped in indigo." And so it appears! 78LY
γ LEO	10 20	+19 51	DS	2.2, 3.5	4"	Algieba. Magnificent, radiant golden suns – one of the finest double stars in the heavens! A 620-year binary. 170LY
54 LEO	10 56	+24 45	DS	4.5, 6.3	6"	Lovely, little-known bluish-white & greenish-white pair. 150LY
R LEO	09 48	+11 26	SS	4.4-10.5	–	Peltier's Variable. Has rosy-scarlet hue throughout its cycle. 600LY
M65/M66/NGC 3628 LEO	11 19	+13 05	GX	9.3/9.0/9.5	10' × 3'/8' × 4'/15 × 4'	Leo Triplet. A trio of bright spirals lying within the same wide field of view – wondrous sight! 30,000,000LY
M95/M96/M105 LEO	10 44	+11 42	GX	9.7/9.2/9.3	7' × 5'/7' × 5'/4' × 4'	Another trio of spirals sharing same field of view! 30,000,000LY
NGC 2903 LEO	09 32	+21 30	GX	8.9	13' × 7'	One of best galaxies missed by Messier – easily spied. 30,000,000LY
γ LEP	05 44	-22 27	DS	3.7, 6.3	96"	Wide pale-yellow & garnet combo "awash in vivid color"! 29LY
R LEP	05 00	-14 48	SS	5.5-11.7	–	Hind's Crimson Star. A gleaming, intense stellar ruby. 1,500LY
M79 LEP	05 24	-24 33	GC	8.0	9'	Winter's Lone Globular. Small & faintish but unique. 50,000LY
α LIB	14 51	-16 02	DS	2.8, 5.2	230"	Zubenelgenubi. Nice wide, binocular & RFT combo. 65LY
Struve 1962 LIB	15 39	-08 47	DS	6.5, 6.6	12"	Pretty, perfectly-matched-white twins – nicely spaced.
ξ LUP	15 57	-33 58	DS	5.3, 5.8	10"	Seldom observed bright, sweet pair of bluish-white suns. 120LY
12 LYN	06 46	+59 27	DS	5.4, 6.0, 7.3	1.7", 9"	Fascinating tight trio, all white. A-B is 700-year binary. 140LY
19 LYN	07 23	+55 17	DS	5.6, 6.5	15"	Attractive pair with subtle contrasting tints. A 9th-mag. lies nearby.
38 LYN	09 19	+36 48	DS	3.9, 6.6	3"	Bright close, unequal pair with elusive color contrast for steady night.

OBJECT/CON	RA	DEC	TYPE	MAG/S	SIZE/SEP	REMARKS
NGC 2419 LYN	07 38	+38 53	GC	10.4	4'	Intergalactic Wanderer. Dim, small & amazingly remote for a globular cluster – 300,000LY!
NGC 2683 LYN	08 53	+33 25	GX	9.7	9' × 2'	Bright, nearly edge-on spiral – cigar shaped. Distance uncertain
α LYR	18 37	+38 47	DS	0.0, 9.5, 9.5	63", 118"	Vega. Dazzling pale-sapphire gem with faint comps. – beautiful! 26LY
β LYR	18 50	+33 22	DS	3.3–4.3, 8.6, 9.9, 9.9	46", 67", 86"	Struve's Eclipsing Binary. Set within starry triangle, forming delicate quadruple. Varies continuously in 13-day period. 860LY
ε-1/2 LYR	18 44	+39 40	DS	5.0, 6.1, 5.2, 5.5	2.6", 2.3"	Famed "Double-Double" multiple system! 600-yr. & 1200-yr. binary pairs 208" apart & slowly orbiting each other. All white. 200LY
ζ LYR	18 45	+37 36	DS	4.3, 5.9	44"	Easy topaz & pale-green double. 155LY
δ LYR	18 54	+36 58	DS	4.5, 5.6	630"	Ultra-wide but lovely reddish-orange & blue-green pair involved in sparse but colorful open cluster Stephenson-1. Both 800LY
Struve 2420 LYR	18 55	+33 58	DS	6.0, 7.7	45"	Fainter & wider miniature of Albireo, lying near the Ring Nebula.
T LYR	18 32	+37 00	SS	7.7–9.6	–	Rather faint but quite stunning! One of the reddest stars known.
M56 LYR	19 17	+30 11	GC	8.2	7'	A dim but sparkling stellar beehive in rich MW field. 45,000LY
M57 LYR	18 54	+33 02	PN	8.8	80" × 60"	Ring Nebula. Finest and best-known planetary in the sky. A celestial smoke ring – superb sight! Central hole visible in small glass. 2,300LY
β MON	06 29	–07 02	DS	4.7, 5.4, 5.6	7", 10"	Herschel's Wonder Star. Striking trio, all bluish-white, forming slender triangle. An amazing spectacle! B-C 3" apart. 700LY
ε = 8 MON	06 24	+04 36	DS	4.5, 6.5	13"	Pretty gold & blue pair in rich Milky Way field.
M50 MON	07 03	–08 20	OC	5.9	16'	Beautiful stellar jewelbox of at least 100 suns. 2,900LY

Object	RA	Dec	Type	Mag	Size	Description
12 MON/ NGC2244/ NGC 22379/ NGC 2246	06 32	+04 52	OC/DN	4.8/–	24'/80' × 60'	Rosette Cluster/Nebula. Huge faint ring-shaped nebulosity surrounding irregular cluster of newborn suns centered on yellow giant. 2,600LY
15 = S MON/NGC 2264	06 41	+09 53	OC	3.9	20'	Christmas Tree Cluster. Big bright cluster of over 40 stars strikingly arranged in the shape of an upside-down evergreen tree! 2,600LY
R MON/NGC 2261	06 39	+08 44	DN	–	2' × 1'	Hubble's Variable Nebula. Small, comet-shaped nebula – changes size, shape & brightness with pulsations of embedded variable. 2,600LY
λ OPH	16 31	+01 59	DS	4.2, 5.2	1.5"	Bright, tight 130-yr. binary. An elongated whitish egg in small glass.
36 OPH	17 15	–26 36	DS	5.1, 5.1, 6.7	5", 730"	Pretty matched close pair with wide comp. All golden-orange. 18LY
o = 39 OPH	17 18	–24 17	DS	5.4, 6.9	10"	Lovely orange & clear-blue jewels – striking!
61 OPH	17 45	+02 35	DS	6.2, 6.6	21"	A neat, nearly-matched duo of silvery-white suns.
70 OPH	18 06	+02 30	DS	4.2, 6.0	5"	Famous yellow & red binary with 88-yr. period. Superb pair! 17LY
M10 OPH	16 57	–04 06	GC	6.6	15'	Big starry ball & near-twin of M12, just 3 degrees apart. 18,000LY
M12 OPH	16 47	–01 57	GC	6.8	15'	Along with M10, the best of the many GCs in OPH. 18,000LY
M14 OPH	17 38	–03 15	GC	7.6	12'	Noticeably fainter but richer cluster than M10 & M12. 33,000LY
M19 OPH	17 03	–26 16	GC	7.2	14'	Oblate Globular. Most oval GC known (from its rapid spin). 30,000LY
M62 OPH	17 01	–30 07	GC	6.6	14'	With M10, brightest GC in OPH. A near-twin of M19. 20,000LY
NGC 6572 OPH	18 12	+06 51	PN	9.0	15" × 12"	Small but intense blue disk like NGC 6210 in HER. 1,900LY
NGC 6633 OPH	18 28	+06 34	OC	4.6	27'	Big, bright scattered clan of nearly 60 stars in unusual shape. 1,000LY

OBJECT/CON	RA	DEC	TYPE	MAG/S	SIZE/SEP	REMARKS
IC 4665 OPH	17 46	+05 43	OC	4.2	41'	Called a Summer Beehive Cluster – sweet in binoculars! 1,300LY
α ORI	05 55	+07 24	SS	0.4–1.3	–	Betelgeuse. Fiery topaz-red supergiant sun – a dazzling gem! 520LY
β ORI	05 14	–08 12	DS	0.1, 6.8	10"	Rigel. Beautiful radiant blue-white supergiant sun with fainter attendant, forming a splendid magnitude-contrast pair! 770LY
η ORI	05 25	–02 24	DS	3.1–3.4, 4.8	1.5"	Bright tight, bluish duo – primary an 8-day eclipsing binary. 1,400LY
λ ORI	05 35	+09 56	DS	3.6, 5.5	4"	Neat cozy pair, both bluish-white with hint of violet or purple. 900LY
δ ORI	05 32	–00 18	DS	1.9–2.1, 6.3	53"	Wide mag.-contrast pair with 5.7-day eclipsing primary. Tints greenish-white & pale-blue or violet. Neat double for binoculars. 1,400LY
ζ ORI	05 41	–01 57	DS	1.9, 4.0	2.5"	Bright close blue-white duo. Flame Neb. (NGC 2024) in field. 1,400LY
23 ORI	05 23	+03 33	DS	5.0, 7.1	32"	Overlooked wide easy combo. Both stars bluish-white in hue.
σ ORI	05 39	–02 36	DS	4.0, 10.3, 7.6, 6.5	11", 13", 43"	Amazing colorful multiple star with faint triple Struve 761 (8.0, 8.5, 9.0, 68", 8") in field; all one vast system! Many hues evident. 1,200LY
ι ORI	05 35	–05 55	DS	2.8, 6.9	11"	Diamond-like pair with Struve 747 (4.8, 5.7, 36") in same radiant gem-field – forming a wide double-double system! 2,000LY
θ-1 ORI	05 35	–05 23	DS	6.4, 7.9, 5.1, 6.7	9", 13", 22"	Famed "Trapezium" multiple star embedded in heart of Orion Nebula. Wondrous spectacle – like diamonds on green velvet! Also several fainter companions – an actual star cluster in formation! 1,600LY
M42/M43 ORI	05 35	–05 23	DN	4.0/9.0	66' × 60'/20' × 10'	Orion Nebula. Finest DN in the sky & perhaps the grandest deep-sky wonder of them all (with the

exception of the MW itself). Magnificent fan-shaped cloud with wings and wisps overflowing the field of view. Obvious emerald-green/turquoise hue with subtle pinkish tints & the Trapezium diamonds at its heart. Thrilling beyond words! 1,600LY

Object	RA	Dec	Type	Mag	Size	Description
θ-2 ORI	05 35	-05 25	DS	5.2, 6.6	52"	Wide bluish-white pair in Orion Nebula.
W ORI	05 05	+01 11	SS	6.2-7.0	–	Its ruddy glow warms the observer on cold Winter nights!
BL ORI	06 26	+14 43	SS	6.3-7.0	–	Another ruby – a twin of W ORI in both hue & brightness.
M78 ORI	05 47	+00 03	DN	8.0	8' × 6'	Weird-looking, comet-shaped nebulosity with two dim stars. 1,400LY
COL 70 ORI	05 36	-01 00	OC	0.4	150'	Epsilon Orionis Cluster. Stunning circular starburst surrounding middle star in Orion's belt as seen in binoculars & RFTs.
ε PEG	21 44	+09 52	DS	2.4, 8.5	143"	Enif/Pendulum Star. Tap scope & see! Yellow & violet. 780LY
M15 PEG	21 30	+12 10	GC	6.4	12'	Rich, compact starball with intense core. 34,000LY
NGC 7331 PEG	22 37	+34 25	GX	9.5	11' × 4'	Big bright, nearly edge-on spiral. 50,000,000LY
η PER	02 51	+55 54	DS	3.8, 8.5	28"	Color & mag.-contrast pair – vivid orange & blue hues. 890LY
α PER/MEL 20	03 22	+49 00	AS	1.8/1.2	185'	Mirfak/Alpha Persei Association. Binocular wonder! 600LY
β PER	03 08	+40 57	SS	2.1-3.4	–	Algol/Demon Star. Naked-eye eclipser – period 2.9 days. 100LY
M34 PER	02 42	+42 47	OC	5.2	35'	"A celestial aegis hung aloft in splendor!" Lovely sight. 1,500LY
M76 PER	01 42	+51 34	PN	11.5	140" × 70"	Little Dumbbell/Barbell/Cork/Butterfly Nebula. Faintish, pearly-white miniature of the Dumbbell Nebula in VUL. 4,000LY

OBJECT/CON	RA	DEC	TYPE	MAG/S	SIZE/SEP	REMARKS
NGC 869/NGC 884 PER	02 19	+57 09	OC/OC	3.5/3.6	30'/30'	Double Cluster. Two magnificent, overlapping radiant starbursts! Amazing colorful, stellar jewelboxes. Awesome in binoculars, RFTs & telescopes of all sizes. Related – 7,200LY & 7,500LY
α PSC	02 02	+02 46	DS	4.2, 5.1	2"	Alrescha. Tight pair with strange subtle tints. 720-yr. binary. 130LY
ψ-1 PSC	01 06	+21 28	DS	5.6, 5.8	30"	Easy matched pair – both stars blue-white.
ζ PSC	01 14	+07 35	DS	5.6, 6.5	23"	Pale-yellow & pale-lilac combo. 140LY
65 PSC	00 50	+27 43	DS	6.3, 6.3	4"	Neatly matched, pale-yellow cozy pair.
TX = 19 PSC	23 46	+03 29	SS	4.5–5.3	–	Lovely reddish-orange sun in Circlet Asterism of PSC. 400LY
α PSA	22 58	–29 37	SS	1.2	–	Fomalhaut. The "Solitary One." A sparkling blue-white gem. 25LY
k PUP	07 39	–26 48	DS	4.5, 4.7	10"	Superb bright pair resembling γ ARI. Both blue-white. 450LY
M46/NGC 2438 PUP	07 42	–14 49	OC/PN	6.1/11.5	27'/66"	Rich uniform clan of over 100 suns with a tiny, ghostly ring-shaped nebula projected against it. Unrelated – 5,400LY & 3,000LY
M47 PUP	07 37	–14 30	OC	4.4	30'	Grand broad splash of several dozen suns. 1,500LY
M93 PUP	07 45	–23 52	OC	6.2	22'	Glorious swarm of some 80 colorful stars. Wedged-shaped. 3,400LY
NGC 2440 PUP	07 42	–18 13	PN	10.5	16"	Tiny, bluish-white disk – a celestial opal. 3,500LY
NGC 2477 PUP	07 52	–38 33	OC	5.8	27'	Superb, rich cluster of 300 stars – like a loose globular. 4,000LY
M71 SGE	19 54	+18 47	GC	8.3	7'	Remote-looking but pretty, misty glow in rich MW field. 13,000LY
AQ SGR	19 34	–16 22	SS	6.7–7.1	–	Glowing reddish stellar ember.
M8/NGC 6530 SGR	18 04	–24 23	DN/OC	5.8/4.6	90' × 40'/15'	Lagoon Nebula. Large floating nebulous patch crossed by great curving dark lane, with scattered cluster to

one side. Wondrous sight! Finest of it class for N. observers after the Orion Nebula. 5,000LY

Name	RA	Dec	Type	Mag	Size	Description
M17 SGR	18 21	-16 11	DN	6.0	46' × 37'	Horseshoe/Omega/Swan Nebula. Multi-named glowing wonder. A long ray with hook at one end, crossed by dark lanes & many stars. 5,000LY
M20 SGR	18 03	-23 02	DN	6.3	29' × 27'	Trifid Nebula. Although inferior to the Lagoon (which lies closeby), a dark-night revelation! Bulbous cloud trisected with dark rifts. 5,500LY
M21 SGR	18 05	-22 30	OC	5.9	13'	Bright stellar clan of some 60 suns lying near the Trifid. 4,000LY
M22 SGR	18 36	-23 54	GC	5.1	24'	M13 Rival. Big, bright magnificent stellar beehive, resolved to center even in small scopes! Stars look ruddy in larger glasses. 10,000LY
M23 SGR	17 57	-19 01	OC	5.5	27'	Big, rich & uniform stellar commune. Lovely sight. 2,100LY
M24 SGR	18 18	-18 25	GX	4.5	120' × 60'	Small Sagittarius Star Cloud. Magnificent MW starcloud for sweeping with binoculars and wide-field telescopes. Overpowering! 16,000LY
M25 SGR	18 32	-19 15	OC	4.6	32'	Large splashy cluster of some 50 suns. Coarse but brilliant. Contains Cepheid U SGR, which varies from 6.3 to 7.1 over 7 days. 2,000LY
M55 SGR	19 40	-30 58	GC	7.0	19'	Large, loosely compressed orb. Needs dark, steady night. 16,000LY
NGC 6818 SGR	19 44	-14 09	PN	9.9	22" × 15"	Little Gem Nebula. Small, bluish-green cosmic egg. 5,000LY
α SCO	16 29	-26 26	DS	0.9–1.8, 5.4	2.5"	Antares. Beautiful fiery-red supergiant with superb emerald-green companion! Very tight – good seeing a must. 900-yr. binary. 600LY

OBJECT/CON	RA	DEC	TYPE	MAG/S	SIZE/SEP	REMARKS
β SCO	16 05	−19 48	DS	2.6, 4.9	14"	Graffias. Lovely blue-white pair resembling Mizar in UMA. 600LY
ν SCO	16 12	−19 28	DS	4.5, 5.3, 6.6, 7.2	1", 2"	Colorful but tight quadruple with pairs 41" apart. Tints subtle but real – striking sight in larger scopes. 440LY
ξ SCO	16 04	−11 22	DS	4.8, 7.3	8"	Yellow pair with Struve 1999 (7.4, 8.1, 12") 280" away forming wide double-double. Primary 46-year period close binary. 80LY
M4 SCO	16 24	−26 32	GC	5.9	26'	Big, softly-shining globular swarm, resolvable in the smallest of scopes. Noticeably elongated vertically. Lovely sight! Near Antares. 7,000LY
M6 SCO	17 40	−32 13	OC	4.2	25'	Butterfly Cluster. Like a butterfly with open wings! 1,400LY
M7 SCO	17 54	−34 39	OC	3.3	80'	Sprawling, radiant swarm of 80 tinted jewels. Binocular target. 800LY
M80 SCO	16 17	−22 59	GC	7.2	9'	Herschel's Delight. Tiny, densely-packed glittering starball. 27,000LY
NGC 6231 SCO	16 54	−41 48	OC	2.6	15'	Glorious, dazzling cluster – 120 suns plus blue supergiant! 6,000LY
NGC 6302 SCO	17 14	−37 06	PN	9.7	2' × 1'	Bug Nebula. Strange, unusual-looking bi-polar nebula. 1,900LY
R SCL	01 27	−32 33	SS	5.9–8.8	–	Pulsating crimson jewel – one of reddest stars in the sky.
NGC 55 SCL	00 15	−39 11	GX	7.9	32' × 6'	Huge, mottled edge-on star-city over ½ degree long. 7,000,000LY
NGC 253 SCL	00 48	−25 17	GX	7.1	25' × 7'	Sculptor Galaxy. Big, bright & beautiful! Cigar-shaped – like a smaller Andromeda Galaxy – a wondrous sight! 7,500,000LY
M11 SCT	18 51	−06 16	OC	5.8	14'	Smyth's Wild Duck Cluster. A rich, glittering fan-shaped swarm of some 500 suns with an 8th-mag. star near apex – a beauty! 5,500LY

Object	RA	Dec	Type	Mag	Size	Description
Milky Way SCT	18 40	−06 00	GX	–	720' × 540'	Scutum Star Cloud/Gem of the Milky Way. "Downtown Milky Way!" An amazing binocular & RFT starry wonderland! Sense 3-D "depth"!!
δ SER	15 35	+10 32	DS	4.2, 5.2	4"	Striking, neatly-paired double with off-white hues – elegant! 85LY
θ SER	18 56	+04 12	DS	4.5, 5.4	22"	Wider version of δ SER. Pretty, easy pair for any glass. 140LY
M5 SER	15 19	+02 05	GC	5.8	17'	M13 Rival. Magnificent ball of stars – a starry blizzard! 25,000LY
M16/IC 4703 SER	18 19	−13 47	OC/DN	6.0/–	25/53' × 28'	Eagle/Star Queen Nebula & Cluster. A faintly fog-bound nebulous star cluster. Site of famous Hubble Space Telescope image. 8,000LY
IC 4756 SER	18 39	+05 27	OC	4.5	70'	Big, bright scattered group of some 80 stars – binocular clan. 1,400LY
NGC 3115 SEX	10 05	−07 43	GX	9.2	8' × 3'	Spindle Galaxy. Elongated glow with bright center – typical elliptical galaxy shape but with pointy ends. 21,000,000LY
118 TAU	05 29	+25 09	DS	5.8, 6.6	5"	Nicely-paired combo – blue-white & bluish. Pretty.
α TAU	04 36	+16 31	SS	0.8–1.0	–	Aldebaran. Lovely topaz gem projected against Hyades Cluster. 65LY
θ-1/2 TAU	04 29	+15 52	DS	3.4, 3.8	337"	Wide naked-eye/binocular pair in Hyades Cluster. White & yellow.
MEL 25 TAU	04 29	+15 52	OC	0.5	330'	Hyades Cluster. Huge bright, striking V-shaped stellar clan abounding in star-pairs & colorful suns. A naked-eye & binocular wonder! 150LY
M1 TAU	05 34	+22 01	SR	8.4	6' × 4'	Rosse's Crab Nebula. Celebrated remnant of the 1054AD supernova outburst with rapidly spinning neutron star/pulsar at core. An irregular pale elliptical glow with ragged edges. Neat close double Struve 742 [7.2, 7.8, 4"] lies unsuspected in field. 6,300LY

OBJECT/CON	RA	DEC	TYPE	MAG/S	SIZE/SEP	REMARKS
M45 TAU	03 47	+24 07	OC	1.2	110'	Pleiades Star Cluster. Brightest, best-known & finest OC in the entire heavens! A brilliant starry commune of blue-white diamonds! Naked-eye, binocular & telescopic wonder. A thrilling spectacle! 410LY
NGC 1514 TAU	04 09	+30 47	PN	10.9	2'	A 9th-mag. star-nucleus surrounded by a faint circular nebulosity. "A most singular phenomenon!" exclaimed Sir William Herschel.
ι TRI	02 12	+30 18	DS	5.3, 6.9	4"	Little-known, close but lovely gold & blue-green pair. 200LY
M33 TRI	01 34	+30 39	GX	5.7	62' × 39'	Pinwheel/Triangulum Galaxy. Big pale, face-on spiral with delicate arms & patches of nebulosity. A dark-night revelation! 3,600,000LY
ζ/80 UMA	13 24	+54 56	DS	2.3, 4.0, 4.0	14", 709"	Famed Mizar with Alcor nearby. Trio of radiant blue-white diamonds! All three suns are spectroscopic binaries (like many other stars on list) & thus one vast sextuple system. First double star discovered. 78LY
ξ UMA	11 18	+31 32	DS	4.3, 4.8	1.8"	Historic 60-year binary (first to have orbit determined) which has made three circuits since discovery! Twin yellowish suns in contact. 26LY
VY UMA	10 45	+67 25	SS	5.9–6.5	–	Ruddy-orange beacon above the Big Dipper – visible year-round.
M81/M82 UMA	09 56	+69 04	GX/GX	6.9/8.4	26' × 14'/11' × 5'	Bode's Nebulae. Finest galaxy pair in sky! M81 is a bright oblong spiral with vivid nucleus; M82 is a long, narrow curved ray crossed by dark rifts. Splendid sight – both floating serenely ½ deg. apart. 7,000,000LY
M97 UMA	11 15	+55 01	PN	11.2	180"	Rosse's Owl Nebula. Large pale nebula with two subtle dark areas or "eyes" making it faintly bi-central. The cigar-shaped 10th-mag. spiral M108 is in the same wide field 48' NW – a true celestial "odd couple"! The

Owl lies 10,000LY away but the galaxy thousands of times as far.

Name	RA	Dec	Type	Mag	Size	Description
M101 UMA	14 03	+54 21	GX	7.7	27' × 26'	Pinwheel Galaxy. Large, pale circular glow – a vast face-on spiral displaying much subtle detail on dark nights. 15,000,000LY
α UMI	02 32	+89 16	DS	1.9–2.1, 9.0	18"	Polaris. Mag.-contrast pair having amazing (apparent) "24-hour orbital period" caused by Earth's rotation! Brightest Cepheid in sky. 430LY
γ VEL	08 10	−47 20	DS	1.8, 4.3	41"	Dazzling bluish pair – one of most beautiful in the heavens! 1,000LY
NGC 3132 VEL	10 08	−40 26	PN	8.2	84" × 52"	Eight-Burst Planetary. One of brightest in sky – white ellipse with 9th-magnitude central sun and hints of multiple rings! 2,000LY
α VIR	13 25	−11 10	SS	0.97	–	Spica. Icy-blue supersun more than 2,000 × Sun's luminosity. 250LY
γ VIR	12 42	−01 27	DS	3.5, 3.5	0.5"	Porrima. Famed bright binary with 171-yr. period. Now opening up from its 2005 minimum separation, these blended stars look like some yellowish cosmic egg with slowly-turning long axis! 39LY
SS VIR	12 25	+00 48	SS	6.0–9.6	–	Ruddy pulsating interstellar beacon – easily spied when at its brightest.
M84/M86/M87 VIR	12 25	+12 53	GX/GX/GX	9.3, 9.2, 8.6	5' × 4'/7' × 6'/7' × 7'	Coma-Virgo Galaxy Cluster. Three bright specimens (all giant elliptical galaxies) of the famed "Realm of the Nebulae." Here, hundreds of star-cities can be seen in small scopes – often several in the same eyepiece field – and more than 10,000 have been photographed! 70,000,000LY
M49 VIR	12 30	+08 00	GX	8.4	9' × 7'	Another bright elliptical positioned between two stars. 65,000,000LY

OBJECT/CON	RA	DEC	TYPE	MAG/S	SIZE/SEP	REMARKS
M59/M60 VIR	12 42	+11 39	GX/GX	9.8/8.8	5' × 3'/7' × 6'	Nice elliptical galaxy pair lying in same field 25' apart.
M61 VIR	12 22	+04 28	GX	9.7	6' × 6'	One of the many spirals in the Coma-Virgo Cluster – face-on with two arms.
M104 VIR	12 40	–11 37	GX	8.3	9' × 4'	Sombrero Galaxy. One of brightest & most spectacular edge-on spirals in the sky! Bulbous glow with dark equatorial band. 28,000,000LY
NGC 4762 VIR	12 53	+11 14	GX	10.2	9' × 2'	The Kite. Thin edge-on like paper kite – dim galaxy NGC 4754 nearby.
3C273 VIR	12 29	+02 03	GX	12.8	–	First Quasar. Also brightest & closest – visible in 4- to 6-inch glass as a dim bluish star despite its vast distance of 1,900,000,000LY!
COL 399 VUL	19 25	+20 11	AS	3.6	60'	Coat Hanger Asterism/Brocchi's Cluster. Like an upside-down starry coat hanger in binoculars. Superb in RFT scopes (which show it erect)!
NGC 6940 VUL	20 35	+28 18	OC	6.3	31'	More than 100 sparkling sapphires – brightest star ruby red! 2,500LY
M27 VUL	20 00	+22 43	PN	7.6	8' × 5'	Dumbbell Nebula. Next to the Ring Nebula, the finest & best-known object of its class! Like a big puffy celestial pillow serenely floating among the stars of the Milky Way Galaxy, where it looks suspended three-dimensionally in space – a truly wondrous spectacle! 1,200LY

About the Author

The author, shown holding a copy of his book *Celestial Harvest: 300-Plus Showpieces of the Heavens for Telescope Viewing & Contemplation*. Originally self-published in 1998 (and updated in 2000), it was reprinted in 2002 by Dover Publications in New York. This labor of love was more than 40 years in the making! Courtesy of Warren Greenwald.

James Mullaney is an astronomy writer, lecturer, and consultant who has published more than 500 articles and five books on observing the wonders of the heavens, and logged over 20,000 hours of stargazing time with the unaided eye, binoculars, and telescopes. Formerly Curator of the Buhl Planetarium and Institute of Popular Science in Pittsburgh and more recently Director of the DuPont Planetarium, he served as staff astronomer at the University of Pittsburgh's

The author shown with his 5-inch Celestron Schmidt–Cassegrain optical-tube assembly mounted on an exquisite old Unitron altazimuth mounting with slow-motion controls. With excellent optics and a total weight of just 12 pounds, this highly portable instrument can go anywhere and is a joy to use. Photo by Sharon Mullaney.

Allegheny Observatory and as an editor for *Sky & Telescope, Astronomy*, and *Star & Sky* magazines. One of the contributors to Carl Sagan's award-winning *Cosmos* PBS television series, he has received recognition for his work from such notables (and fellow stargazers) as Sir Arthur Clarke, Johnny Carson, Ray Bradbury, Dr Wernher von Braun, and his former student, NASA scientist/astronaut Dr Jay Abt. His 50-year mission as a "celestial evangelist" has been to "Celebrate the Universe!" – to get others to look up at the majesty of the night sky and to experience personally the joys of stargazing. In February, 2005, he was elected a Fellow of the prestigious Royal Astronomical Society (London).

Index

Unless mentioned or illustrated in the text, individual celestial objects are not listed here. Deep-sky wonders (including stars) can be readily found under their respective constellations in Appendix 3.